BEING, NATURE, AND LIFE IN ARISTOTLE

This volume of essays explores major connected themes in Aristotle's metaphysics, philosophy of nature, and ethics, especially themes related to essence, definition, teleology, activity, potentiality, and the highest good. The volume is united by the belief that all aspects of Aristotle's work need to be studied together if any one of the areas of thought is to be fully understood. Many of the chapters were contributions to a conference at the University of Pittsburgh entitled "Being, Nature, and Life in Aristotle," to honor Professor Allan Gotthelf's many contributions to the field of ancient philosophy; a few are contributions from those who were invited but could not attend. The contributors, all longstanding friends of Professor Gotthelf, are among the most accomplished scholars in the field of ancient philosophy today.

JAMES G. LENNOX is Professor of History and Philosophy of Science at the University of Pittsburgh. He has published many articles in professional journals and edited volumes, and his most recent books are *Aristotle:* On the Parts of Animals, *Translation with Introduction and Commentary* (2001), and *Aristotle's Philosophy of Biology: Studies in the Origins of Life Science* (Cambridge, 2001).

ROBERT BOLTON is Professor of Philosophy at Rutgers University. He is author or editor of numerous books and articles on topics in ancient philosophy, particularly those concerning questions in epistemology and philosophy of mind, and those concerning philosophical and scientific methods and their applications in natural science, metaphysics, and ethics.

BEING, NATURE, AND LIFE IN ARISTOTLE

Essays in Honor of Allan Gotthelf

EDITED BY

JAMES G. LENNOX AND ROBERT BOLTON

CAMBRIDGE
UNIVERSITY PRESS

CAMBRIDGE UNIVERSITY PRESS
Cambridge, New York, Melbourne, Madrid, Cape Town, Singapore,
São Paulo, Delhi, Dubai, Tokyo, Mexico City

Cambridge University Press
The Edinburgh Building, Cambridge CB2 8RU, UK

Published in the United States of America by Cambridge University Press, New York

www.cambridge.org
Information on this title: www.cambridge.org/9780521768443

First published 2010

Printed in the United Kingdom at the University Press, Cambridge

A catalog record for this publication is available from the British Library

Library of Congress Cataloging in Publication data
Being, nature, and life in Aristotle : essays in honor of Allan Gotthelf /
edited by James G. Lennox, Robert Bolton.
p. cm.
Includes bibliographical references and index.
ISBN 978-0-521-76844-3 (hardback)
1. Aristotle. 2. Metaphysics. 3. Ontology. 4. Gotthelf, Allan, 1942–
I. Lennox, James G. II. Bolton, Robert (Robert A. N.)
B485.B383 2010
185 – dc22 2010035490

ISBN 978-0-521-76844-3 Hardback

Contents

Preface

A number of the chapters in this volume began as presentations at a conference celebrating the many contributions of Allan Gotthelf to the study of ancient philosophy that took place on October 1–3, 2004. It gives the editors great pleasure to thank the many individuals and organizations who contributed to the success of that event. At the University of Pittsburgh both financial and organizational support were provided by the Center for Philosophy of Science and its staff, Karen Kovalchick, Joyce MacDonald, and Carol Weber. Further financial support was provided by the University of Pittsburgh's School of Arts and Sciences, and Departments of History and Philosophy of Science and Philosophy, and by the Rutgers University Endowment for Ancient Philosophy. Alec Stewart, Dean of the Honors College of the University of Pittsburgh kindly made its remarkable rooms on the thirty-fifth and thirty-sixth stories of the Cathedral of Learning available for a gala reception, funding for which was provided by the Anthem Foundation for Objectivist Scholarship.

We would like to thank Michael Sharp for his support and patience in working with us at all stages on the production of the present volume; Cambridge University Press's referees, whose suggestions helped to improve the final product; Elizabeth Hanlon and Sarah Roberts for their careful handling of editorial matters; Jan Chapman for her skillful copy-editing; Benjamin Goldberg for help in preparing the indexes; and finally, the Syndics of Cambridge University Press for approving this volume for publication.

Notes on contributors

ROBERT BOLTON is Professor of Philosophy at Rutgers University. He is author of *Science, dialectic et ethique chez Aristote* (2010), and co-editor of *Form, Matter and Mixture in Aristotle* (1996) and of *Logic, Dialectic and Science in Aristotle* (1994). He is a former Rhodes Scholar and Fellow of the Centre National de la Recherche Scientifique, Paris.

SARAH BROADIE is Professor of Philosophy at the University of St. Andrews. She has also taught at the University of Edinburgh, the University of Texas at Austin, Yale, Rutgers, and Princeton. She specializes in Ancient Greek Philosophy and her publications include *Aristotle and Beyond* (Cambridge, 2005); *Aristotle: the* Nicomachean Ethics (with Christopher Rowe) (2002); *Ethics with Aristotle* (1991); *Nature, Change and Agency in Aristotle's* Physics (1982); and *Passage and Possibility, a Study of Aristotle's Modal Concepts* (1982).

DAVID CHARLES is Research Professor of Philosophy at Oriel College, Oxford. He is the author of *Aristotle on Meaning and Essence* (2000), *Aristotle's Philosophy of Action* (1984) and the editor of *The Greeks on Definition* (2010).

ALAN CODE is a Board of Governors Professor of Philosophy at Rutgers University; he was previously the Nicholas C. Petris Professor of Greek Studies at the University of California at Berkeley, and the O'Donnell Professor of Philosophy at the Ohio State University. His research has concentrated on Aristotle's metaphysics and logic, and related topics in his natural philosophy and biology.

JOHN M. COOPER is Henry Putnam University Professor of Philosophy at Princeton University, and Director of the joint Ph.D. Program in Classical Philosophy there. He is editor of *Plato: Complete Works* (1997), co-author with John Procopé of *Seneca: Moral and Political Essays* (1995), and author of *Reason and Human Good in Aristotle* (1975), *Reason and*

Emotion: Essays on Ancient Moral Psychology and Ethical Theory (1999), and *Knowledge, Nature, and the Good: Essays on Ancient Philosophy* (2004).

MARY LOUISE GILL is Professor of Philosophy and Classics at Brown University. She is the author of *Aristotle on Substance: The Paradox of Unity* (1989), co-translator and author of the Introduction, *Plato:* Parmenides (1996), and co-editor of *Unity, Identity, and Explanation in Aristotle's* Metaphysics (1994), *Self-Motion from Aristotle to Newton* (1994), and *A Companion to Ancient Philosophy* (2006). She is currently completing a book entitled Philosophos: *Plato's Missing Dialogue.*

ARYEH KOSMAN is the John Whitehead Professor of Philosophy Emeritus at Haverford College, where he began teaching in 1962, after education at the University of California, Berkeley, Hebrew University, and Harvard. He specializes in the history of philosophy, particularly ancient philosophy, but continues to follow his earliest interests in contemporary analytic philosophy. He has held visiting positions at the University of California, Berkeley, UCLA, the University of Washington, Princeton, and the University of Pittsburgh.

JAMES G. LENNOX is Professor of History and Philosophy of Science and Director of the Classics, Philosophy and Ancient Science Program, University of Pittsburgh. He is the author of a translation, with introduction and commentary, of *Aristotle:* On the Parts of Animals (2001); and of *Aristotle's Philosophy of Biology: Studies in the Origins of Life Science* (Cambridge, 2001). He has held fellowships at the Center for Hellenic Studies (1983–4); Clare Hall, University of Cambridge (1986–7); the Istituto di Studi Avanzati, University of Bologna (2006).

PIERRE PELLEGRIN is Director of Research at the Centre National de la Recherche Scientifique (CNRS) in Paris and former director of the Center for the History of Science and of Arab and Medieval Philosophy. He has worked primarily on Aristotle's natural philosophy, notably his biology (*Aristotle's Classification of Animals: Biology and the Conceptual Unity of the Aristotelian Corpus*, 1986) and on Aristotle's political philosophy. He also specializes in ancient biology generally, and in Greek and Roman skepticism. He has translated works of Aristotle, Galen, and Sextus Empiricus into French.

DAVID SEDLEY is Laurence Professor of Ancient Philosophy at the University of Cambridge, and a Fellow of Christ's College. His books include

The Hellenistic Philosophers (1987, with A. A. Long), and *Creationism and its Critics in Antiquity* (2007), based on his 2004 Sather Lectures. He edited *Oxford Studies in Ancient Philosophy* from 1998 to 2007. He is a Fellow of the British Academy, and a Foreign Honorary Member of the American Academy of Arts and Sciences.

Acknowledgments

Parts of Chapter 1 appeared as Chapter 6 of *Creationism and its Critics in Antiquity*, University of California Press: Berkeley, Los Angeles, and London, 2007.

An earlier version of Chapter 5 (substantially revised and expanded for the present volume) appeared as "The Unity of Substances in *Metaphysics* H.6" (in Portuguese), in *Cadernos de História e Filosofia da Ciência*, series 3, 13 (2003): 177–203.

Earlier versions of Chapters 8 and 9 were published in French in *Dunamis: Autour de la puissance chez Aristote*, ed. M. Crubellier, A. Jaulin, and D. Lefebvre, Peeters: Leuven, 2008.

An earlier version of Chapter 10 has been translated into Portuguese and will appear in *Sobre a Ética de Aristóteles – textos selecionados*, ed. Marco Zingano, Odysseus Editores: São Paulo, Brazil, 2010.

Allan Gotthelf: a biographical sketch

Allan Gotthelf was born on December 30, 1942, in Brooklyn, New York, the first of two children of Max and Dorothy Gotthelf. He thus came of age during the glory years of the Brooklyn Dodgers – he recalls attending games at Ebbets Field, once watching Jackie Robinson (who joined the Dodgers' roster in 1947) steal home, and forming an informal 'Gil Hodges Fan Club' with two friends. But as passionate as he was for sports, his true love was understanding things at the deepest level, and after doing three years of junior high school in two he attended the justly famous Stuyvesant High School, with its rigorous training in mathematics and science, from 1956 to 1959. (Stuyvesant has graduated an astonishing number of accomplished alumni, including four Nobel laureates – perhaps Joshua Lederberg being the most well known – and actors ranging from James Cagney to Lucy Liu.)

Prior to discovering philosophy, Allan focused his thirst for understanding on mathematics and physics, and in 1959, at the age of sixteen, he entered Brooklyn College, intending to major in physics but shifting after he arrived toward theoretical mathematics. During the summer of 1961 he read Ayn Rand's *Atlas Shrugged*, which influenced him to redirect his intellectual focus toward philosophy. He graduated in 1963 with a Major in Mathematics and a Minor in Philosophy, having taken classes in philosophy with Martin Lean and John Hospers. Though he had decided to pursue an advanced degree in philosophy, he had a strong interest in philosophy of mathematics and had already accepted a graduate assistantship at Penn State University in mathematics. So after completing his MA in mathematics there in one year, he entered the graduate program in philosophy at Columbia University in 1964.

Over the next two years he completed his course work and then spent three years as a full time instructor, at Wesleyan University. He eventually settled on "Aristotle's Conception of Final Causality" as the topic for his dissertation, and received his Ph.D. in 1975. An essay based on his dissertation won the dissertation essay prize of the *Review of Metaphysics* and was published in its 1976/7 volume. By then he had already received tenure in the Department of Philosophy and Religion at Trenton State College (now The College of New Jersey [TCNJ]) and during his years there he was instrumental in fostering a thriving philosophy club and in the development of the Honors College, and a special Minor in Classical Studies. He chaired the department from 1988 to 1997, during which time the Philosophy Major grew fourfold. He was passionate about teaching philosophy, playing an active role in the American Philosophical Association (APA) Teaching Workshops. From 1982 to 1990 he and Michael Hooker were selected as co-leaders of annual weekend workshops in the Eastern United States.

In the twenty years between 1980 and 2000, Allan played a central role in organizing conferences, workshops, and summer institutes that encouraged scholars of Aristotle's philosophy to integrate the study of Aristotle's biological works into their research. Many of the contributors to the present volume, including its editors, were participants in the first of these, a conference organized in collaboration with David Balme that took place during the summer of 1983 at Williams College in Williamstown, MA. Allan had befriended David Balme during the latter's visit to the Center for Advanced Study in Princeton in 1976 and interacted constantly with him until Balme's untimely death in 1989.

Other collaborations on related themes followed, with Sir Geoffrey Lloyd, Michael Frede, John Cooper, Pierre Pellegrin, and Wolfgang Kullmann (a detailed list is provided on pp. 268–9), and Allan gained a reputation in the community of Aristotle scholars as a prodigious organizer. A number of these conferences led to the publications mentioned at the beginning of our introduction, publications credited with moving Aristotle's biological writings to the center of Aristotle scholarship.

During his years at TCNJ Allan held a number of visiting positions and fellowships: at Swarthmore in 1974–5; at the Center for Hellenic Studies in 1982/3; at Oxford (where he co-taught a seminar with John Ackrill), and at Clare Hall, Cambridge, in 1984; and at Georgetown in the spring of 1985. Having first gotten to know Gotthelf during his visit to Oxford in 1984 (and again during a number of visits to Oxford and to Clare Hall, Cambridge (where he had been appointed Life Member in 1985), David Charles invited him back in 1994 to co-teach a seminar with him and one of this volume's editors (Lennox) called "Philosophical Issues in Aristotle's Biology." Shortly after leaving Oxford he spent five weeks in Japan, presenting a week long intensive course at Toyko Metropolitan

University entitled "Aristotle's Biological Enterprise and its Philosophical Significance" and lecturing in five other cities, including Kyoto.

As noted earlier, Allan developed a close friendship with David Balme, widely considered the leading scholar of the twentieth century on Aristotle's biological works. In fact his first edited volume was a festschrift in honor of Balme, *Aristotle on Nature and Living Things: Philosophical and Historical Studies*, published in 1985. Allan worked collaboratively with Balme during the 1970s and 80s, and when Balme died 1989, he was invited by the Loeb Classical Library to see Balme's Loeb edition of *History of Animals, Books VII–X* through to publication; and then, working closely with the Balme family, he took on the much more daunting task of preparing Balme's draft of his planned *editio maior* of the entire *History of Animals* for publication as a two-volume work in the *Cambridge Classical Texts and Commentaries* series. Volume I appeared in 2002; Volume II is currently projected for publication in 2011. Gotthelf took early retirement from TCNJ in 2002, at which time the College created the Gotthelf Prize to be awarded annually to an outstanding graduating senior at TCNJ, who is chosen by the Classical Studies Faculty.

Upon retiring, Allan accepted an offer to be a visiting professor for the fall term at University of Texas, Austin, and since 2003, thanks to the creation of a Fellowship for the Study of Objectivism in the Department of History and Philosophy of Science (HPS), University of Pittsburgh, he has been Visiting Professor of HPS at Pittsburgh.

Earlier it was noted that his turn to philosophy from mathematics and science, and to Aristotle in particular, was due to the influence of Ayn Rand. Allan was among a number of young philosophers with whom Rand met regularly to discuss her more recently developed philosophical work, and throughout his career Allan has devoted the same energy and focus as he had devoted to putting Aristotle's biological works "on the map," to putting Objectivism, Rand's philosophy, on the contemporary philosophical map. He has been a prime mover behind the Ayn Rand Society, which became affiliated with the APA Eastern Division in 1988, and his fellowship at the University of Pittsburgh is appropriately designed to support work both on Aristotle and Rand, and on the relationship between Rand's philosophy and Aristotle's. Since coming to Pittsburgh he has organized a number of workshops on both philosophers, and a number of informal reading groups; he has taught and co-taught seminars on Aristotle's biological works and served on dissertation committees of students working on Aristotle. One serendipitous consequence of his joining the faculty in Pittsburgh was that he was able to serve on the dissertation committee of his most promising undergraduate at TCNJ, Greg Salmieri, who joined Pittsburgh's graduate program in philosophy a year before his former teacher took up his Fellowship in HPS. Allan is hopeful that within a year of this collection of essays in his honor appearing in print, a collection of his papers, including some not previous published, scheduled for publication by Oxford in their Oxford Aristotle Studies series, will also appear.

Note on abbreviations

Abbreviations of ancient works correspond to those used in H. G. Liddell, R. Scott, and H. S. Jones, *A Greek–English Lexicon*, 9th edn., Oxford, 1996.

Introduction

The chapters of this volume represent some of the most promising results of current advanced research on a number of related themes in the thought of Aristotle, and in so doing aim to honor the many and varied contributions of Allan Gotthelf to Aristotelian studies. These two aims are not accidentally connected. Perhaps no one has done more in recent decades to promote, to sponsor, to organize, and to stimulate research on these topics in Aristotle's thought than Allan Gotthelf. The explorations and ideas on display here all reflect and, to a significant degree, derive from his efforts. As such, they constitute a fitting token of esteem from his friends and colleagues. As he would be the first to agree, nothing could be more suitable, as a way of honoring him and his accomplishments, than to further advance our understanding of Aristotle.

Many of the essays in this collection were first presented at a gathering held at the University of Pittsburgh on October 1–3, 2004 under the rubric: "Being, Nature, and Life: A Conference Celebrating Allan Gotthelf's Contributions to the Study of Classical Philosophy and Science." Others are contributed by scholars who were unable to attend. The main themes here are those of the conference: Aristotle's metaphysics, his natural science and biology, and his methodology. All of these studies exhibit, to a greater or lesser extent, the interconnections among these topics in Aristotle's own thinking.

To appreciate the significance of these studies, and their place in recent and continuing work on Aristotle, it will be useful to refer back to an earlier collection, conceived and edited by Gotthelf himself jointly with one of the current editors, advice from the other, and contributions from both. That collection, *Philosophical Issues in Aristotle's Biology* (Cambridge, 1987), brought new attention to, and displayed the importance of, the material found in Aristotle's biological works for our understanding of his doctrines in the same central areas of his thought mainly in focus here – his philosophical and scientific method, his natural philosophy with its teleological focus, and his metaphysics. That collection itself was preceded by an initial conference dealing largely with these same topics, primarily organized and directed by Allan Gotthelf, which was held in Williamstown, Massachusetts in the summer of 1983, a conference in which many of the current contributors participated. Subsequent collections, deriving from later conferences in whose organization Allan Gotthelf again played a leading role, drew on this earlier work and further advanced the discussion of these and related themes.[1]

While the authors of the current studies have not intentionally looked back to the material in that original collection or its successors, these essays do belong in numerous ways to a still continuing stream of research which in no small part stems from those earlier endeavors. Indeed, as we have noted, in most of these studies attention is given to more than one of the three main areas of Aristotle's thought investigated in the original collection. This can be seen at once in the opening chapter, by David Sedley, which serves to introduce a major overarching theme linking all of the essays in this volume: Aristotle's natural teleology. Sedley's study draws not only on Aristotle's natural science proper but also on features of Aristotle's scientific method, concerning particularly his views on causal explanation, and also on special material from Aristotle's metaphysics. It is the perceived cross-fertilization in Aristotle's investigation of these different areas, against the background of Plato's reflections on the same themes, that leads Sedley to his main results.

In a similar vein, Robert Bolton's chapter on the relation between Aristotle's biology and his metaphysics uses material both from the *Analytics* and the *Metaphysics* to delineate the boundary line which Aristotle fixes between metaphysics and biology, and it investigates the implications of this for central doctrines in both areas of his thought. In a group of related studies James Lennox, Alan Code, and Mary Louise Gill pay special attention to a distinctive methodological device that plays an important role not only in Aristotle's procedures of inquiry in natural science but also in metaphysics – his method of definition by division. Lennox argues for a narrative unity behind the structure of *On the Parts of Animals* I and identifies a primary role for the method of division in the early stages of inquiry in natural science, prior to the search for causes. Code reaches a similar result mainly through an independent comparative study of material in the *Posterior Analytics* and *Metaphysics* concerning Aristotle's theory of definition. Gill argues for another distinctive role for the method of

[1] See Devereux and Pellegrin 1990; and Kullmann and Föllinger 1997.

division in Aristotle's treatment of a special problem in metaphysics, the unity of substance. A comparison of her study with those of Lennox and Code points to a common interest on Aristotle's part in the use of division both in natural science and in metaphysics in order to fix on proper kinds and proper data for study. So in this group of chapters Aristotle's interests in method can again be seen to mesh closely with his scientific concerns, both in biology and metaphysics. Pierre Pellegrin also explores Aristotle's theory of definition, by way of a close investigation of book 2 of the *Posterior Analytics*. His focus is on Aristotle's account of how ultimate causal definitions, and demonstrative explanations based on them, are reached in science.

Aryeh Kosman, David Charles, and Sarah Broadie offer a further group of connected chapters, in which detailed attention is given to a main feature of Aristotle's metaphysics, namely his fundamental, and closely connected, distinctions between form and matter and between actuality and potentiality, as these figure both in his biology and and in his metaphysics. Kosman's chapter explores particularly, in this connection, the relation between formal and efficient causes in Aristotle's account of biological generation, while those of Charles and Broadie explore the relationship between actuality (or activity), potentiality and teleology in the context of *Metaphysics* Θ.

In a final, complementary chapter John Cooper explores the topic of Aristotle's teleology in Aristotle's ethics and political philosophy, reminding us that this motif not only figures prominently in Aristotle's science and philosophy of science but also in his approach to the question of the highest good for human beings. Cooper's study reveals how strongly the uses of teleology in natural and metaphysical science – uses displayed in nearly all of the earlier essays – are echoed in Aristotle's account of distinctively human life and human good.

As will be apparent, we have not thought it appropriate to try to summarize in any detail here the contents of the particular essays included below. Each makes highly distinctive and carefully argued claims whose intricacies invite and mandate close study by the reader. We have rather tried to indicate how, taken together, these chapters form a certain whole. A careful reader will in fact discern many more interconnections among these essays than we have tried to describe here.

Seen, then, from the perspective of the tradition of scholarship to which these studies belong, this collection is more than the sum of its parts. It advances a certain multi-faceted common research project in which all of the contributors have been engaged for more than twenty-five years now, under early and continuing stimulus from Allan Gotthelf, a project whose underlying assumption is that Aristotle's philosophical and scientific method, his natural science and biology, his metaphysics, and his moral and political theory too, need to be studied together if any one of these areas of his thought is to be fully understood.

CHAPTER I

Teleology, Aristotelian and Platonic

David Sedley

GOD AS PARADIGM

Aristotle was Plato's student for two decades before founding his own school. Is it more fruitful to think of his mature work as anti-Platonist, or as that of an independent Platonist? Although this age-old question does not admit of final resolution, I am convinced with regard to my present topic, the explanation of purposive structures in the world,[1] that most can be learnt by emphasizing, rather than minimizing, Aristotle's Platonic background and training.[2]

Plato, like nearly every other thinker in and well after antiquity, associated teleology with conscious purpose. To make the world a purposive structure just is to posit an intelligent mind as its cause. True, the intelligent mind could have created the world and then left it to run itself mechanically, but no ancient thinker – after at any rate Anaxagoras, whose position on the point is open to dispute – was ready to contemplate a split-level theory of that kind. Either the world was intelligently created and is intelligently run, or it originated from non-intelligent causes and is still, with the possible exception of human action, governed by causes of that same kind. While the atomists defended the latter view, Plato developed the former:

This chapter is based on material from my book *Creationism and its Critics in Antiquity* (Sedley 2007), and I am grateful to the University of California Press for permission to reuse it here. In working on the topic, I have received help from people too numerous to list in this note, although they are listed, as exhaustively as I can manage, in the book's preface. In addition, Errol Katayama and Larry Jost have been kind enough to supply criticisms of the book that have prompted me to make some changes to the current version. Let me finally say what it pleasure it is to be able to dedicate the chapter to my friend Allan Gotthelf, whose work on Aristotelian teleology has been truly seminal.

[1] I must here leave untouched many of the major issues in Aristotle's teleology, on which see esp. Gotthelf 1997c.

[2] Gerson 2005, provocatively entitled *Aristotle and Other Platonists*, should be consulted for a much more ambitious, and more Neoplatonic, assimilation of the two than I have contemplated, including chapter 4 on issues relating to causation.

his Demiurge, who created the world, has left it under the overall control of the intelligent and divine world-soul.[3]

In conformity to this background, Aristotle too treats the twin issues of creation and administration in strict parallel to each other. The world, along with its resident species, is not the product of an intelligent act of creation, for the simple reason that it had no beginning at all but has always existed – a thesis he defends by appeal to the essential eternity of the heaven's circular motion. And likewise when it comes to the world's continued functioning, there is no divine oversight, planning or enforcement.[4] So far he may seem to tend closer to the atomist camp, since no divine interest in our world is invoked at any stage. But like Plato, and unlike the atomists, he nevertheless holds that throughout the natural world there are irreducibly purposive structures. Pretty well everything in nature has a purpose, despite the fact that no intelligence either conceived that purpose or administers it.

This restrained teleology has won Aristotle innumerable admirers. For, it is rightly said, purposive structures are indeed basic to nature, quite regardless of the question of divine control or its absence. Never mind whether you are a creationist or the most hardened of Darwinians: you cannot avoid saying that the heart is for pumping blood, the eyelid for protecting the eye, the teeth for cutting and grinding food. Nor, for the Darwinian, are these locutions just a shorthand for some more accurate mode of biological explanation: adequate non-teleological explanations of the parts of the eye are simply not available.

Now, it is one thing to commend Aristotle for the refreshing modernity of his teleological thinking, and to contrast it with the antiquated creationism of a Plato. It is quite another thing to suppose that the outlook's appearance of modernity is the key to Aristotle's own rationale in developing the theory in this particular way. Ancient atomism likewise resulted in a great many modern-looking theses, yet it started from premises utterly unlike those of modern or even early modern physics. Similarly, or so I shall argue, Aristotle's minimalist approach to purpose in nature is very far from being a sign of his modernity. Such a recognition should not, however, lessen our appreciation of the light that his ideas can be used to shed.

Where did the motivation for Aristotle's revised teleology come from? From an unexpected quarter, it seems. Plato had famously conceded in

[3] It is admittedly hard to establish how much more the Platonic world-soul governs than the celestial rotations. But *Ti.* 37a4–c5 (on which cf. Reydams-Schils 1997) makes it clear that it has true 'opinions' (δόξαι) about the sensible world of becoming and hence does not concentrate its thought exclusively on pure being.

[4] True, Aristotle does occasionally talk as if god can be credited with some providential action, e.g. *GC* 2.10.336b27–34, but I join the consensus that regards such locutions as merely figurative (see esp. Solmsen 1963, pp. 485–95, but cf. Bodéüs 2000 for a less dismissive reading).

book 7 of the *Republic* that, for a philosopher, government ranks second best to the life of pure contemplation. And correspondingly in the *Timaeus*, where he maintains that the entire world has been so structured as to enable the rational human soul to imitate the divine mind through the study of astronomy and philosophy, this imitation of god is located in pure mathematical and philosophical contemplation, rather than in the exercise of moral or political virtues.[5] Yet the divine creator and divine world-soul of Plato's *Timaeus* are themselves viewed as partly engaged in world-management. If so, they are not, even on Plato's own view, exclusively engaged in the best activity available to them.

Aristotle is in this regard more Platonist than Plato himself. He too (*EN* 10.7–8) holds that the kind of happiness that can come from leading a virtuous civic life, although of great value, is second-best to the godlike happiness of pure detached contemplation. But he improves on Plato to the extent that he seeks to make his own theology consistent with that same ranking of different brands of happiness. God's activity can only be the best, he argues in *Metaphysics* Λ, and, if so, it must be pure contemplation.[6]

The effect of this minor-looking adjustment to Platonism is breathtakingly far-reaching. If god must be a pure contemplator, he cannot be an administrator.[7] There can therefore be no Demiurge, and no divine world-soul. In which case, the world is uncreated and functions without divine oversight. The outcome is, in short, Aristotle's cosmology.

In positing a detached and self-absorbed god, one who is above any inclination to intervene in our world, Aristotle sounds surprisingly similar to Epicurus. Yet unlike Epicurus he fully shares with Plato the conviction that god is the supreme explanatory principle. And he reconciles these two apparently conflicting motifs – god as detached and god as causally supreme – by drawing on another Platonic idea: that god is the supreme object of emulation. The goal of life, as Plato's followers expressed his idea, is 'to become as like god as possible'.[8] Plato meant this goal mainly as a human aspiration, although in two contexts (*Smp.* 207c9–208b6; *Lg.* 4.721b6–c8) he extended it to the entire animal kingdom by presenting the drive to propagate as mortal organisms' best shot at achieving

5 I defend this interpretation of the *Timaeus* in Sedley 2000a.

6 It is entirely possible that this complete insulation of god from practical activity took time to evolve and is not yet fully worked out in the *On the Heavens*, although even there note, for example, 2.12.292a22–8 on god as free from *praxis*.

7 For an especially forthright development of this theology on Aristotle's behalf, cf. the passages of Alexander of Aphrodisias cited and discussed by Sharples 2003.

8 Annas 1999 chapter 3; Sedley 2000a.

immortality – surrogate immortality, that is, achieved by living on through their descendants – and the perpetuity of their species. Aristotle takes up this latter cue and develops the idea still further. The supreme divinity is an unmoved mover, a detached self-contemplator, whose activity is pure actuality, and *everything* else in the world functions by striving, in its own way, to emulate that actuality.[9]

The highest human aspiration, philosophical contemplation, is the most direct imitation of god's own activity (*EN* 10.7.1177b26–1178a8; 10.8.1178b7–32). Procreation, in humans, lower animals and plants is, as it had been for Plato, a bid for immortality by proxy, another way of imitating god's eternal actuality, namely by perpetuating both oneself and one's species (*de An.* 2.4.415a26–b7; *Pol.* 1.2.1252a28–30; *GA* 2.1.731b24–732a1; *Metaph.* Θ.8.1050b28–30).[10] Even below the level of plant life, the world's natural cycles, such as the weather cycle whereby the four elementary bodies undergo endlessly repeated intertransformations, are imitations of god's eternal actuality (*Mete.* 1.9.346b35–347a10; *GC* 2.10.336b34–337a7).

It is, in short, scarcely an exaggeration to say that for Aristotle the entire functioning of the natural world, as also that of the heavens, is ultimately to be understood as a shared striving towards godlike actuality.[11] Admittedly Aristotle does not very often stand back to view the matter panoramically in this way, for his interest is far more often taken up with specific biological structures and processes and their contribution to the organism's success; but he does view it along these lines in *Metaphysics* Λ.10, the culminating chapter of his theological treatise, to which I shall return in the final section of this chapter.

Even biological structures that might have been accounted for in far more down-to-earth ways are, on occasion, brought by Aristotle under the same explanatory principle of striving for godlikeness. According to Plato in the *Timaeus* (45a3–b2), describing the original creation of the human body, our creators made the face the natural front (here, as in Aristotle's biology, defined by the orientation of the senses), because front is 'more honourable' (*timiōteron*) than back, being both of higher ranking and more appropriate to leadership. This evaluative ranking of directions, which the modernizing interpretation would happily have seen Aristotle consign to history, is a doctrine which on the contrary he develops and frequently

[9] Cf. *EE* 7.15.1249b13–15, 'For it is not by giving commands that god is ruler, but as the good towards which practical wisdom gives commands.'

[10] Cf. Burnyeat 2004, 24 for the possibility that in addition the latter cycles are for the sake of the former.

[11] See esp. Kahn 1985.

exploits. According to Aristotle, not only is front more honourable than back, but so are right than left and up than down.

To take the up–down polarity, man's unique uprightness makes him superior, to the extent that his natural up, his head, coincides with the cosmos's upward direction (*PA* 2.10.656a7–13):

> Of this kind is the human race. For it, either alone or most of all among animals known to us, shares in the divine . . . It is alone in having its natural parts in the natural arrangement, and its up is related in the natural way to the universe's up. For alone of the animals, man is upright.[12]

At the other end of the biological spectrum, plants have their natural 'up' coinciding with the cosmic down, in that their roots – which are functionally their mouths – are down in the soil. For example (*de An.* 2.4.416a2–5):

> For up and down are not the same for each and every being but as the head is in animals so the roots are in plants, if one ought to call organs different or the same by their functions.

Virtually all of this is Platonic in origin. Compare *Timaeus* 90a2–b1:

> Concerning the most authoritative kind of soul found in us, we must have the following thought. God has given it to each of us as a daimon – this thing which we say dwells at the topmost part of our body and raises us up from the earth towards what is akin to us in the heaven, because we rightly call ourselves a heavenly plant, not an earthly one. For the divinity keeps our body upright by suspending our head and root from the place out of which our soul was first born.

We can here see how Aristotle's treatment of plants as inverted human beings has its origin in Plato's elevation of human beings to the status of inverted plants. In so far as Aristotle gives 'up' and 'down' their own specialized biological senses – as Plato had already done in speaking of a 'natural front' – Aristotle is no doubt saying something scientifically credible.[13] Try drinking a cup of tea while standing on your head. We might intelligibly ask if you can get the tea to go 'down' your throat. Even though its geographical or cosmic direction of flow is up, we understand the expression because in context we assign 'down' a meaning determined by biological function alone. Aristotle similarly assigns directions like up

[12] There are many other relevant passages in Aristotle, but I here limit myself to citing for comparison *HA* 1.15.494a20–b1 (which supplies a good deal more detail), and *Juv.* 477a21–3. Gregoric 2005 is a valuable comparison of Plato's and Aristotle's explanations of human erect posture, but in my view it does not pay sufficient attention to the intrinsic value that both attach to our postural kinship with cosmic topography, for which (regarding Plato) cf. Osborne 1988, 107–9.

[13] Thus Lennox 1985, 149–55, replying to Lloyd 1966, 52–61. It will be evident that my main sympathies are with Lloyd, but Lennox is a valuable guide to the controversy.

and down their own functional biological senses (*Progression of Animals* [*IA*] 4–5).

Nevertheless, the use to which Aristotle puts this mode of expression is hardly so innocuous. When he presents up, front and right as 'better and more honourable' (*PA* 3.3.665a6–26), he does so in the context of explaining why, in human anatomy, nature sometimes favours these orientations even when practical utility alone would not. Thus the heart, as governing organ, occupies the 'honourable' front,[14] even though that in turn requires the windpipe, which services it, to be 'badly placed' in front of the oesophagus; the resultant danger of food going down the wrong hole then requires the remedial provision of the epiglottis.

The reason why, regardless of practical utility, Aristotle attaches an honourable status to the body's sharing the cosmic inclinations up, front and right is his belief, also derived from Plato,[15] that these orientations represent the divine source or direction of motion governing the rotation of the heavens themselves (*Cael.* 2.2). The fact that in the anatomy of the best animals, namely humans, inclinations to front, up and right are repeatedly favoured is therefore one expression of our superior likeness to the divine heavens, and through them to god.

If the whole natural world is, in one way or another, pulling itself up by its own bootstraps in the interests of maximum godlikeness, how is that possible? Desire is a faculty that, according to Aristotle, is found only in animals, yet he is explicit that plants too strive for immortality through reproduction, and that in some attenuated way even the four elementary bodies strive for everlasting actuality. Almost certainly the notion of striving will have to be interpreted reductively, as describing an inherent natural tendency. Such psychologizing descriptions of non-psychological processes, misleading and indispensable in equal measure, have been commonplace in the history of science, from 'Nature abhors a vacuum' to Natural Selection and the Selfish Gene. Certainly the *Timaeus* is full of them, including intelligence's 'persuasion' of the four elementary stuffs.

But even after effecting such a reduction, we are left with the following result. The reason why in Aristotle's view no directive mind can be at work in natural processes is not any preference on his part for 'scientific' over theological modes of explanation. It lies rather in the conviction that the

[14] Unlike Lennox 1985, 149–52 (cf. Lennox 2001a, 254), I understand *PA* 3.3.665a19–26 as saying that the reason why the heart must necessarily be at the front is that it is more honourable, and not that it is practically advantageous.

[15] In the *Timaeus* heavenly rotation is both to the right, 36c6, and forwards, 40b1. Cf. also 90a2–b1, quoted above.

Platonic account gets the theology *wrong*. God's causality in the natural world is omnipresent, as Plato held, but must be such that all the operative drives and impulses belong to the natural entities, leaving god himself eternally detached and self-focused.

In the remainder of this chapter I shall consider two other areas in which Aristotle's teleology seems to me to reflect his Platonic beginnings. One is the fundamental role of the craft model in his accounts of nature; the other is his assumption – which surfaces more frequently than is generally appreciated – that purpose is present in nature not just at the level of individual organisms, but also in the entire world's interactive structure and functioning.

THE CRAFT ANALOGY

Plato had appealed not merely to divine benevolence in explaining the natural world, but also to divine craft, which was itself in turn to be understood in terms of familiar human craft. It could hardly be denied that Plato had been stunningly successful in explaining the natural world as the product of craftsmanship, and Aristotle must have been loath to throw away those gains. Yet divine craftsmanship was ruled out by his theology, as we have seen. No natural process could be acceptably explained, along the lines set by Plato, as decreed by the deliberations of a world-soul or of any other immanent deity, let alone a transcendent one. Such then was Aristotle's dilemma. He resolved it, as *Physics* 2 eloquently attests, by developing his conviction that although nature is not divine craft, it is sufficiently *analogous* to craft in its working for much of the light shed by Plato to remain. So wedded is he to the craft analogy that its characteristic language is ubiquitous in his biological writings, where he frequently, for example, speaks figuratively of nature as an agent 'crafting' (*dēmiourgein*, cognate with 'Demiurge') her products.[16]

Aristotle's whole understanding of natural processes relies on his famous distinction between four kinds of cause. And one prominent way in which the craft model comes to his aid is in enabling him to elucidate that very distinction. His preferred methodology is, as he reminds us at the opening of the *Physics* (1.1.184a16–21), to start with what is more familiar, or makes more sense, *to us* (that which is *gnōrimōteron hēmin*), and to move from there to what makes more sense *in its own right* (that which is *gnōrimōteron*

[16] Cooper 1982, although to my mind a most penetrating portrayal of Aristotle's teleology, is I think wrong to deny the craft analogy's centrality to Aristotle's physical thought. For a convincing reply on this point, see Broadie 1990, 392–6.

haplōs or *phusei*). The causal processes of nature make more sense in their own right and exhibit teleology in a higher degree, but the causal processes of craft make more sense to us, because all of us have practised a craft or witnessed one at close quarters.

Aristotle therefore uses a miscellany of craft-related examples to demarcate the four causes (*Ph.* 2.3), before moving on to their application to nature. In a craft, it is normal for the practitioner to impose some form on pre-existing matter, for example a sculptor on bronze. Here the *material* cause is the bronze, the *moving* cause (often called the 'efficient' cause) is the sculptor, and the *formal* cause is the form he imposes on it, all unmistakably distinct from each other. In addition to these, the sculptor works with a goal in view, perhaps the completion or perfection of the statue, and this too is a distinct motivating factor in the story, the *final* (meaning 'end-related') cause.[17] Thus four distinct causes – material, moving, formal and final – are clearly exhibited by craft, and it is only once we have mastered them and their interrelation in this familiar guise that we are ready to look for them in nature.

In nature, the four causes are much trickier to disentangle from each other. Take the development of a pig, and to begin with, its material cause. The pig's matter – its flesh, bones, and the like – is certainly a causal factor helping to constitute its nature but is not nearly as readily distinguishable from the pig's form as the bronze was from the shape of the statue it constituted; for one thing, there never was a time in the pig's history or prehistory when all of these specific materials existed without already having the form of pig, in the way that before the statue was made the bronze was there to be inspected. For an analogous reason the pig's formal cause, that is, its essential form as a pig, is not at first sight fully distinct from its matter. Third, take the moving cause of the changes the pig undergoes. This lies initially (according to Aristotle's theory of animal generation) in its father and hence is external to it, as the sculptor was to the bronze, but during its lifetime its active moving cause becomes an internal one, since this role of mover is taken over by the pig's own soul, identifiable once more with its essential form. Nor is the final cause of its growth, that is, the end or goal governing the process, any more straightforwardly distinguishable

[17] Admittedly this full set of four causes of a statue, although it becomes common in the subsequent tradition (see Todd 1976), occurs nowhere in Aristotle. The examples used in *Ph.* 2.3 to illustrate the four kinds of cause and their further taxonomy are a miscellany that includes, in addition to the sculptor and his bronze, the arts of building, navigation, musicianship, spelling, syllogizing and medicine, as well as some altogether non-technical fields.

from its form, because that goal really just *is* its fully developed form as a pig, towards which it is striving.[18]

Nature, then, is difficult. But if we start from the causal distinctions that craft clarifies, we can aspire to understand it. And there is a further reason why we should hope to progress from craft to nature. A craft, according to Aristotle, is an extension of what nature already does. A craft takes over where nature leaves off, imitating and completing nature's work. For this very reason, he argues (*Ph.* 2.8.199a15–20), we can work out that nature already embodies goals or purposes, because that is the only possible source of the goals which crafts adopt. For instance, since medicine is a craft that aims to help the body regain its health, that goal of regaining health can be seen to have already been governing the internal natural healing processes, which the doctor intervenes merely to facilitate and complete.

At the top end of the *scala naturae*, nature forms a near-continuum with craft, it being at times hard to tell where the one stops and the other starts. How, then, does craft still *differ* from nature? This is a vital question. For, once Aristotle has identified the residual differences between craft and nature, his project is to devise perspectives and thought experiments which will shrink or even erase those differences. Such is his procedure for obtaining illumination from the craft–nature analogy, while stopping short of Plato's position, that of fully identifying nature with divine craft.

The first and most obvious difference between craft and nature, in Aristotle's eyes, is that in craft the moving cause is regularly external to the matter. Carpenters are external to their wood, cooks are external to their ingredients, doctors are external to their patients. It is precisely for this reason that the crafts teach us, more clearly than nature can, that the moving cause acting upon the matter really is a distinct factor in the process. Yet even in craft the moving cause is not *essentially* external, Aristotle points out, and here the gap between craft and nature can be narrowed: if a doctor treats himself, the change in his own body from illness to health results from an internal moving cause. That is *very* much like a natural process. Admittedly in this special case of the self-curing doctor the moving cause is not *essentially* internal, as it would be in nature. It just happens that the doctor is also the patient, and it could easily have been otherwise (*Ph.*2.1.192b23–7). Still, as Aristotle illuminatingly says, when a doctor cures himself this is just about as near to nature as craft can come (2.8.199b30–2).

[18] E.g. *Ph.* 2.7.198a24–7; *PA* 1.1.641a25–7.

Perhaps Aristotle should have left it at that. But, to take a yet further step in the same direction, he asks us to perform a thought experiment (199b28–30): we are invited to imagine the craft of shipbuilding as being present not in the shipbuilder, but in the timber itself. If that were so, and the moving cause had become a source of change internal to the matter instead of external to it, we would see ships grow naturally out of piles of planks. This particular image may sound too bizarre to be helpful, and it is a pity that Aristotle did not instead choose for his example a material which could more easily be imagined as shaping itself from inside, such as a lump of clay growing into a pot, the potter's craft having been implanted in it instead of imposed from outside. Even then what such thought experiments reveal, and conceivably are designed to concede, is that the craft–nature gap could never be closed altogether. Provided that they illuminate the continuum of purpose between artificial and natural processes, they have done their job.

But is not the kind of purposiveness that is present in nature exactly where this analogy fails? The most obvious of objections to Aristotle's reasoning is that purpose is present in craft precisely because of the involvement of a human craftsman with an intellect. And is not it his most central thesis that this craftsmanship has no analogue in nature, god being entirely detached from the natural world? Without the admission of a creative or productive intellect operative in natural processes, how can purpose exist in nature? And how can the craft analogy be of any assistance?

Here is Aristotle's enigmatic remark on the question (199b26–8):

> It is ridiculous for people not to believe that something is coming about for a purpose if they do not see that the moving cause has deliberated. Yet craft too does not deliberate.[19]

Craft does not deliberate? But of course it does, at least in our everyday experience of crafts on which Aristotle's methodology leans so heavily. Scholars have, in my view, generated unnecessary difficulties over the interpretation of this admittedly dark remark. It is regularly suggested that Aristotle has in mind an idealized picture of the craftsman. This ideal practitioner's skill has developed to such a high degree that he no longer has to ask himself how to perform a particular operation, that is, to deliberate: he just goes ahead and does it.[20]

[19] ἄτοπον δὲ τὸ μὴ οἴεσθαι ἕνεκά του γίγνεσθαι, ἐὰν μὴ ἴδωσι τὸ κινοῦν βουλευσάμενον. καίτοι καὶ ἡ τέχνη οὐ βουλεύεται.

[20] The best defence of this interpretation known to me is Broadie 1990. I am grateful to Sarah Broadie for correspondence on this and related issues.

The trouble is that to the best of my knowledge nowhere else in the Aristotelian corpus has such an idealization of craft yet been found.[21] Quite apart from the fact that few of us would put our lives in the hands of a surgeon who did not deliberate before amputating a leg, or of an airline pilot who did not choose a flight path before take-off, Aristotle himself again and again depicts the operations of craft precisely in terms of deliberation.[22] Deliberation is how we choose the best means to a given end, and the presence of deliberation in a process is therefore a salient sign that it is end-directed. However, for reasons we have yet to discover, Aristotle says in the quoted passage that crafts do not in fact deliberate.

For a more satisfactory understanding of this puzzling remark, I am convinced that we must go back to Aristotle's causal theory. In it, his debt to Plato is greater than is sometimes recognized. For Plato, a fundamental causal process is one which is analytically self-evident.[23] For example, heat makes you hot, and wisdom makes you wise. Such statements are hardly informative, and in modern usage they would not normally be deemed causal at all, but at least they seem analytically true, and for Plato that is where their great merit lies. Other, more obviously synthetic causal statements, such as that jogging makes you hot, or that studying geometry makes you wise, are more debatable precisely because they do not keep the cause–effect relation itself transparent.

Now Aristotle's ultimate commitment to this Platonic causal principle is discernible in many contexts, but none more so than *Physics* 2.3, the famous chapter in which he formally expounds his theory of four causes, and which constitutes the immediate background to chapter 8, where the puzzling remark about craft is located. One example that he repeatedly invokes there is that of the moving cause of a building (*oikia*). To answer

[21] Robert Bolton (this volume, Chapter 2) draws attention to *Metaph.* 1.1.981b2–5, a passage in which it might be thought to be said that craftsmen act by habit rather than deliberation. I take this passage to be rather about the procedures of a mere menial worker (χειροτέχνης), who falls well below the level of a craftsman (τεχνίτης). The craftsman applies his knowledge of the relevant causal processes (981a24–30), whereas the menial worker has no such knowledge and simply follows the instructions of his site-director (ἀρχιτέκτων), thus leaving no room for deliberation.

[22] Cf. n. 25 below. True, in his account of deliberation at *EN* 3.3.1112a34–b8 Aristotle allows that in certain relatively precise disciplines, such as spelling, deliberation is superfluous. But in that same context he is careful to call these disciplines 'sciences' (ἐπιστῆμαι), and contrasts them with 'crafts' (τέχναι), including medicine, in which he insists that deliberation typically *is* required. What is presumably meant to be the same distinction between deliberative medicine and non-deliberative spelling is made at *EE* 2.10.1226a33–b2 (I thank Errol Katayama for the reference), this time not in terms of ἐπιστήμη vs τέχνη but of the lack of any need for reasoning when spelling (we must remember that Greek was spelt phonetically). Importantly, medicine is his favoured craft in the *Physics* when it comes to the craft–nature analogy.

[23] See further, Sedley 1991.

the question 'What is this moving cause of this building (*oikia*)?' by saying that it is a builder (*oikodomos*) is an excellent answer. Why? Because the causal relationship between builder and building is far from accidental, in that builders and buildings are suitably related much in the way that wisdom, as cause, is suitably related to the effect of someone's being wise. But suppose that the builder is called Jones. To have said, instead, that the moving cause of the building is Jones would, although true, have stated a merely accidental cause. There is nothing about his being Jones that links him causally or explanatorily to the building, in the way that his being a builder manifestly does.

Nevertheless, even to refer to the moving cause of the building as 'a builder' falls short of the ideal answer, it turns out. Near the end of chapter 3, Aristotle writes (195b21–5):

> One must always seek the ultimate cause of each thing, as in other matters. For example, a man builds because he is a builder, but the builder builds in virtue of the building craft. This cause, therefore, is prior, and the same applies in all cases.[24]

That is, even when we nominate a builder as the cause of the building we have not gone all the way to isolating its immediate moving cause. Strictly speaking, the building craft, located in the builder's soul, is the ultimate cause that accounts for the building. Jones builds in virtue of being a builder, and the builder builds in virtue of the building craft. We cannot continue this regress, because there is no further thing in virtue of which the building craft causes the building. With the building craft, therefore, we have found the ultimate moving cause.

This admittedly unintuitive refocusing of our causal language serves a vital purpose in Aristotle's metaphysics. Form, he holds, is eternal. In nature, an organism's form pre-exists it, typically by being already present in its father, who according to Aristotle was its original and external moving cause. (Try substituting the thought that your genetic code in some sense pre-existed you in your ancestors.) To throw light on this, Aristotle compares the pre-existence of form in nature to the way that in a craft the artefact's form pre-exists the artefact itself, by being already present in the mind of the craftsman. The form of some actual building, for example, existed before it was built, namely in the mind of its builder. Thus he writes in *Metaphysics* Z.7 (1032b5–14):

[24] δεῖ δ' ἀεὶ τὸ αἴτιον ἑκάστου τὸ ἀκρότατον ζητεῖν, ὥσπερ καὶ ἐπὶ τῶν ἄλλων. οἷον ἄνθρωπος οἰκοδομεῖ ὅτι οἰκοδόμος, ὁ δ' οἰκοδόμος κατὰ τὴν οἰκοδομικήν· τοῦτο τοίνυν πρότερον τὸ αἴτιον, καὶ οὕτως ἐπὶ πάντων.

Health is the formula in the soul, and the science. The healthy state comes to be when the doctor has thought as follows: 'Since such and such is health, if a healthy state is to be, such and such must exist, for example a balanced condition, and if this is going to be, heat.' And so he goes on thinking until he brings it down to an end point which he can himself enact.[25] Thereafter, the movement towards a healthy condition which starts from this is already called production. Hence it results that in a way health comes to be from health, and building from building, namely the material building from an immaterial building. For the medical craft and the building craft are, respectively, the form of health and the form of the building. What I am calling immaterial being is the thing's essence.

Vitally, the building craft *is* the immaterial form or essence of the building, resident in the builder's soul before he imposes that same essential form on the bricks and mortar.

In the light of all this, we can return to the enigmatic pronouncement of *Physics* 2.8, 'It is ridiculous for people not to believe that something is coming about for the sake of something if they do not see that the moving cause has deliberated. Yet craft too does not deliberate.' Aristotle does not mean to deny that the craftsman deliberates. But the craftsman is not, in the strictest sense, the moving cause. The ultimate moving cause is, as we have seen, the craft itself, identifiable with the essential form of the product resident from the beginning in the craftsman's soul. And that ultimate moving cause does not do any deliberating.

Seen in this light, Aristotle's strategy is not, as often thought, to deny that deliberation is on the one hand present in crafts but on the other hand absent from nature. His point is rather that, when you strip down to its hard core the causality by which in each of the two domains the moving cause operates, the deliberation that occurs in craft becomes a strictly ancillary factor. In craft and nature alike, an essential form serves as a moving cause which brings about its own imposition on the relevant matter. The form of the building, present initially in the builder's soul, prompts the movements which end in that same form's being fully present in the bricks and mortar. The form of pig, present originally in the piglet's father and later progressively in the piglet itself, prompts the movements which end in that same form's being fully realized in the mature adult pig.

The use of analogy, such as Aristotle's craft–nature analogy, unavoidably requires that some gap remain between the two compared items. At the same time, however, the greater the number of differences between the

[25] Notice how Aristotle develops this point without for a moment wanting to minimize the role of conscious deliberation in the process by which the form is transferred to the external matter. For a builder's deliberation, cf. also *PA* 1 1.639b25–30.

analogically related items that can be eliminated or marginalized, the more persuasive and informative the analogy becomes. Differences undoubtedly remain between the two processes, and the fact that deliberation plays a part in craft but not in natural processes is one of these. But such differences in Aristotle's eyes should not be allowed to mask the underlying isomorphism between the two causal processes. And pointing out that in neither case does the ultimate moving cause, namely the essential form, do any thinking helps to confirm how deep that isomorphism runs. This is why the causal structure of craft really does enlighten us about the causal structure of nature.

GLOBAL TELEOLOGY

Metaphysics Λ.10 is the vital culminating chapter of Aristotle's theology, and in it he speaks fairly explicitly of a global teleology, attributing cosmic goodness to what he calls 'the nature of the whole' (1075a11–25):

> We must consider also in which way the nature of the whole (ἡ τοῦ ὅλου φύσις) possesses the good and the best – whether as something separated and by itself, or as its arrangement. Or is it in both ways, like an army? For an army's goodness is in its ordering and is also the general. And more the general, since he is not owing to the arrangement, but the arrangement is owing to him. All things are in some joint arrangement, but not in the same way – even creatures that swim, creatures that fly, and plants. And the arrangement is not such that one thing has no relation to another. They do have a relation: for all things are jointly arranged in relation to one thing. But it is like in a household, where the free have least licence to act as they chance to, but all or most of what they do is arranged, while the slaves and beasts can do a little towards what is communal but act mostly as they chance to. For that is the kind of principle that nature is of each of them (τοιαύτη γὰρ ἑκάστου ἀρχὴ αὐτῶν ἡ φύσις ἐστίν). I mean, for example, that at least each of them must necessarily come to be dissolved; and there are likewise other things in which all share towards the whole.

This passage has been unwelcome to Aristotelian scholars who deny that – following Plato's lead – he advocates a global teleology, and some have sought to show that no such teleology is intended. Just what structures he might mean to include in this cosmic nature, if such he intends, is uncertain, and not a topic on which I plan to dwell at length here, but the broader context leaves no doubt that it starts with the divine Unmoved Mover and the celestial movements it inspires, and that it extends down to terrestrial ecology, with a particular emphasis on the kinds of movement manifested at the level of animals.[26] My aim in this final section is not to explore the cosmological detail, but rather to provide confirmatory evidence that Aristotle really does think there is, in addition to individual natures, a global nature, causally governed by the Unmoved Mover. This global nature brings with it a global teleology, and here once again Aristotle is closer to his Platonic roots than is generally conceded.

I shall start this time from *Physics* 2.4–6, and from Aristotle's denial there, in response to the atomists, that luck could ever account for the structure of the world taken as a whole. The atomists' position on this is set out, with some derision, in chapter 4 (196a24–35):

> Some people consider the fortuitous the cause of this heaven and of all the worlds, explaining that it was fortuitously that there arose the vortex and the motion which separated things and set the universe in this arrangement. This is pretty amazing. For on the one hand they say that animals and plants neither are nor come to be by luck, but that either nature or intelligence or some other such thing is their cause (it not being just anything that arises from each seed, but an olive tree from this one, a man from that one), yet on the other hand they say that the heaven and the most divine of perceptible things came to be fortuitously, without any cause comparable to that of animals and plants.

Aristotle here finds it inconsistent of materialists to recognize the role of purposive processes and structures within our world, yet to deny an analogous cause of the vastly superior and more ordered[27] structure of the world itself. When he speaks of the atomists admitting that animals and plants are the products of 'nature or intelligence or some other such thing', he can hardly be talking about the origin of species, which he knows they would have vehemently denied to be the work of a creative intelligence. He must rather be invoking their agreement that individual animals and plants today come into being, not by luck, but either through natural propagation, or, as implied by his example of olive trees, thanks

[26] See fuller discussion in Sedley 2000b, where I argue in particular that the second reference to 'nature' (retaining unemended the MS reading τοιαύτη γὰρ ἑκάστου ἀρχὴ αὐτῶν ἡ φύσις ἐστίν at 1075a22–3) enables us to recognize a second reference to global nature. Bodnár 2005, 18–19, is right, I think, to reply that the sentence *could* still be read as referring to individual nature. But it becomes much the less natural reading, because 'the nature of the whole' is the announced topic. My preferred reading also gives the ensuing clause (on which Bodnár does not comment) much more point: that each organism is eventually dissolved is hardly an obvious example of the kind of principle *its* nature is but nicely illustrates the eternal recycling of the elements that is part of the global teleology (above, p. 8), with a distant echo of Plato's assertion that the matter we are made of has been 'borrowed' from the world (*Ti.* 42e6–43a1), implying that it must be duly returned.

[27] This point about greater orderliness is made explicit in the version of the argument at *PA* 1.1.641b10–23, discussed below.

to the intelligent operations of farming.[28] This is meant to show that they perfectly well understand the notion of luck and correctly avoid applying it to familiar purposive processes, yet they misapply it when it comes to the origin of the cosmos.

By the close of chapter 6, Aristotle has developed his own account of luck (*tuchē*) and the fortuitous (*to automaton*),[29] which for present purposes I shall not keep distinct. Luck is an accidental moving cause. As I understand him, the lucky outcome always has a *per se* cause: for example, collecting a debt has as its *per se* cause the antecedent desire, or need, to collect the debt. When you go to the marketplace to buy bread and happen to see your debtor, your desire to buy bread accidentally coincides with that desire or need. This reapplies the causal principles we met above (pp. 15–17): when a flute-player builds a building, the flute-player accidentally coincides with the builder, so that the moving cause of the building is *per se* a builder, and only accidentally a flute-player. Builder and building are (as their very names reveal) suitably correlated as cause and effect, whereas flute-player and building are not. Likewise the desire to collect the debt is suitably correlated, as cause, to collecting the debt, as effect, whereas the desire to buy bread is not.

At the end of his excursus on the fortuitous (*Ph.* 2.4–6), Aristotle returns to this theme of the cosmos itself (198a5–13):

Since the fortuitous and luck are causes of things of which intelligence or nature *could* be the cause, whenever something accidentally becomes the cause of these same things, and since nothing accidental is prior to things that are *per se*, it is clear that the accidental cause is not prior to the *per se* cause either. Hence the fortuitous and luck are posterior to intelligence and nature. So however much it might be true that the fortuitous is the cause of the heaven, it is necessary that intelligence and nature are prior causes, both of many other things and, especially, of this universe.[30]

Aristotle means the following. Lucky or fortuitous events are, on his analysis, events that lead to the accidental fulfilment of a pre-existing natural

[28] See Wardy 2005 for the complications surrounding the question to what extent Aristotle can comfortably regard olive trees and other cultivated varieties as either 'natural' or 'artificial'. It seems likely, nevertheless, that at least for dialectical purposes he is here treating the propagation of the olive as artificial (and hence the effect of *nous*), that of man as natural.

[29] I gratefully borrow this translation of τὸ αὐτόματον from Sharples 1983.

[30] ἐπεὶ δ' ἐστὶ τὸ αὐτόματον καὶ ἡ τύχη αἴτια ὧν ἂν ἢ νοῦς γένοιτο αἴτιος ἢ φύσις, ὅταν κατὰ συμβεβηκὸς αἴτιόν τι γένηται τούτων αὐτῶν, οὐδὲν δὲ κατὰ συμβεβηκός ἐστι πρότερον τῶν καθ' αὑτό, δῆλον ὅτι οὐδὲ τὸ κατὰ συμβεβηκὸς αἴτιον πρότερον τοῦ καθ' αὑτό. ὕστερον ἄρα τὸ αὐτόματον καὶ ἡ τύχη καὶ νοῦ καὶ φύσεως. ὥστ' εἰ ὅτι μάλιστα τοῦ οὐρανοῦ αἴτιον τὸ αὐτόματον, ἀνάγκη πρότερον νοῦν αἴτιον καὶ φύσιν εἶναι καὶ ἄλλων πολλῶν καὶ τοῦδε τοῦ παντός.

or psychological goal. Hence, even if our world were assumed to have had an origin, and luck were hypothesized to have played a part in that origin, this would still amount to a concession that the goal accidentally fulfilled by it was already operative before luck intervened to bring it to fruition.

This concluding argument is an easily overlooked but important indication as to how we are to understand Aristotle's foregoing notion of luck. It might have been – and has been – thought that when Aristotle makes luck a moving cause which accidentally coincides with the *per se* cause, the *per se* cause need not actually exist.[31] But the present argument would make no sense unless he were assuming that, on the contrary, the *per se* cause must be an actual one. Your desire or need to collect the debt may not have been operative before you saw your debtor, but, once you saw him, that *per se* cause kicked in and led to its own proper effect, the completion of the transaction. Or, even supposing it did not, it was a sufficiently actual presence in the story to explain why we call it 'lucky' that you saw your debtor and collected the debt: without there actually being that antecedent desire or need, there would have been nothing lucky about it at all. Likewise, Aristotle is now telling us, there would be nothing 'lucky' about the world's coming into being unless there had been an antecedent moving cause of the world, whether that cause be intelligence or nature.

One might respond that some lucky events produce desirable results that were neither desired nor needed in advance. For example, in the Chinese fable recounted by Charles Lamb, when a house complete with its pigsty burnt down and roast pork was accidentally discovered,[32] prior to the lucky event there was neither a desire nor a need for roast pork. Aristotle will have to reply that, if the event is to count as lucky at all, it must have fulfilled an antecedent goal under *some* description: if not the goal of eating roast pork, at least that of eating nutritious or tasty food. If no such goal had existed, under any suitable description, what could be counted lucky about the event? Likewise for the world's origin (assuming for the moment that it had one), we cannot intelligibly invoke the fortuitous or luck as cause without assuming that, under *some* description, an antecedent goal was being fulfilled by it. And as soon as the atomists concede that, they have collapsed back into the correct position that they were earlier represented

[31] This may be the one major point on which I part company with the excellent analysis of Aristotelian chance in Judson 1991. As I understand him (esp. p. 92), Judson sees chance events as having no operative *per se* cause at all. This might indeed have been what Aristotle meant, but the concluding argument of 2.6 seems to me to exclude it.

[32] Lamb 1823. Prior to that they had eaten their pork raw. The humorous twist to the tale is that thereafter, whenever they wanted roast pork, they would burn the house down.

as unreasonably resisting. That is, they will be treating the world's origin as on a par with the regular regeneration of life forms within our world, which even hardened materialists such as they are must admit is not purely fortuitous.

On this hypothesis that the world had a beginning,[33] when Aristotle says that *intelligence and nature* are prior to any luck that might have been involved, he means priority in a temporal sense: an intelligent and natural goal must have already existed at a time before the world's creation. But his own preferred view, outside this adversarial context, is that the world in fact had no beginning. On that hypothesis, what does he mean by saying that intelligence and nature are causes of the world prior to any luck that might be involved? For the point of the argument would be lost if these were not causes he intended here to advocate in his own voice.

This time the priority must be causal or explanatory, rather than temporal. The world is an eternal purposive structure, causally dependent on intelligence and nature. The 'intelligence' (*nous*) he intends cannot be a Platonic Demiurge: instead, he can only be referring to that detached divine intellect the Prime Mover, which as the ultimate source of all change he does indeed consider causally prior to the world, and, as the ultimate model for emulation, the *per se* cause of all the changes lower down the scale. But he has also concluded that *nature* is a prior cause in relation to the world. The world taken as a whole, he must mean, is no accidental structure, but manifests natural purpose just as much as individual organisms do.

This reading is confirmed by a closely parallel text, taken from a key methodological chapter, *On the Parts of Animals* 1.1 (641b10–23):

Moreover, nothing abstract can be an object studied by physics, because nature does everything for a purpose. For just as in artefacts art is present, in things themselves there appears another such principle and cause, *which, like the hot and the cold, we have from the universe* (ἣν ἔχομεν καθάπερ τὸ θερμὸν καὶ τὸ ψυχρὸν ἐκ τοῦ παντός). Hence it is more reasonable for the heaven to have come to be by the agency of such a cause, if it has come to be, and to be because of such a cause, than mortal animals. At any rate, what is ordered and definite is much more evident in celestial things than as belonging to us, whereas what belongs more to mortal beings is variability over time, and the fortuitous. Some say that although each animal is and comes to be by nature, the heaven has been formed in the way it is from luck and the fortuitous. Yet nothing whatsoever in the heaven appears the result of luck and disorder.

[33] For this as a hypothesis that Aristotle takes seriously enough to work with in a context like the present one, cf. *PA* 1.1.641b10–23, quoted below on this page.

Here the closing critique of the atomists[34] plainly corresponds to the one we have just met in *Physics* 2.4 and 2.6. It differs in arguing just that the *heaven* cannot be the product of luck or the fortuitous, and not this time explicitly extending that conclusion to the causation of the entire world, but we need not doubt that the two passages share a common agenda and are to be read in the light of each other.

In leading up to his refutation of the atomists, Aristotle's second sentence supplies a premise that was not explicit in the other passage. Final causality, he asserts here, is primarily present in the world as a whole, and in beings like us only derivatively, just as the hot and the cold in us are derivative from cosmic hot and cold. This inference is an unnoticed (as far as I am aware) direct descendant of the classic argument placed in the mouth of Socrates by Xenophon (*Mem.* 1.4.8) and Plato (*Phlb.* 29a9–30d9).[35] It occurs in a chapter of Xenophon frequently echoed by Aristotle in his biological writings:[36]

> Do you think that there is nothing wise anywhere else? And this despite your knowing that there is a lot of earth, of which you have just a little portion in your body, a lot of moisture, of which you have just a tiny bit, and that each of the other things too is huge whereas your body has been fitted together by acquiring a tiny portion of them? Yet intelligence alone, it turns out, you think exists nowhere, and that you laid your hand on it by some stroke of luck? And that it is due to some kind of unintelligence that these vast masses of limitless amount are well arranged?

Just as the water, earth and suchlike in our bodies are derived from the great cosmic masses of these same stuffs, Socrates argues here, so too our intelligence must be inferred to be derivative from a great cosmic intelligence which is responsible for the world's good ordering.

Aristotle's version replaces water and earth with two of his own elementary powers, the hot and the cold, and also replaces intelligence with final causality, in conformity with his by now familiar modification of Platonic teleology to remove conscious purpose from nature. But his argument is unmistakably in the same tradition.[37] Aristotle's prioritization here of

[34] Although I agree with Lennox 2001a, 136, that *PA* 1.1.640b4–17 is not specifically targeted at Democritus, it seems overcautious (pp. 136 and 145) to extend the same doubt to the present passage or to its doublet in *Ph.* 2.4.

[35] The same Socratic argument was later elaborately reworked by the Stoics as the very basis of their theology: see Sedley 2005.

[36] E.g. compare *Mem.* 1.4.6 on the arrangement of teeth, with Aristotle, *Ph.* 2.8.198b24–6, *PA* 3.1.661b6–9, and, on eyebrows and eyelashes, with *PA* 2.15.658b14–26.

[37] Johnson 2005, 136–7 with n. 9, translates ἔχομεν . . . ἐκ τοῦ παντός not 'we have from the universe' (as I do above, along with all other translators I have consulted, including now Lennox 2001a), but 'we ascertain from the universe', describing the sentence as 'obscure', as it would indeed be, thus translated.

cosmic teleology over that of individual natural processes[38] is a vital but under-appreciated motif, to which I now return.

Clearly the world as a whole is structured in many ways that are regularly beneficial to life, including the availability of natural resources, and the eternally recurrent intertransformation of the four simple bodies that underlies the weather cycle, this latter dependent in turn on the daily and annual cycles of the sun. If Aristotle conceded that these advantageous cosmic structures require no teleological explanation, he would be playing into the hands of his opponents by implying that advantageous structures in individual organisms might equally well be understood as non-purposive.[39] Worse still from his point of view, he would be making the teleologically governed perpetuation of species in the sublunary world dependent on an overall cosmic arrangement whose own eternity was itself not explicable in terms of purpose but just a matter of brute fact. It is easy to see why Aristotle wants to avoid any such dependence of the teleological on the non-teleological, and why instead we have just seen him, on the contrary, emphasizing that biological teleology is derivative from a prior cosmic teleology.

Just such a derivation had been a prominent feature of the key background text, Plato's *Timaeus*. There, the world's goodness is the primary explanandum, and that same world's comprehensive stocking with organic species, each elaborately engineered to suit its specific function, is in turn explained by its contribution to that goal. Readers of Aristotle have had difficulty in believing that he can have inherited any such prioritization from his teacher, because his zoological treatises rarely mention global teleology[40] and instead concentrate on the functioning of individual organisms in their own right. But we should not be misled by this emphasis, because it too has its background in Plato. Although the *Timaeus*, taken as a whole, leaves no possible doubt about the subservience of zoological to global teleology, it is its long treatments of anatomy at 69c5–86a8 and 90e1–92c3 that correspond functionally to Aristotelian zoology, and these sections have very

[38] This prioritization reflects Aristotle's pronouncements (*Mete.* 1.1.338a20–b22, 339a5–9) on the correct order of study, with celestial motion preceding zoology and botany. See further, Burnyeat 2004, 13–24; Falcon 2005, 2–13.

[39] Indeed, not just equally well, but *a fortiori*, since Aristotle here makes sublunary teleology dependent upon its cosmic counterpart. This prioritization does not, however, mean that an organism is so structured as to put its cosmic role before its own self-interest (as Pellegrin 2002 and Johnson 2005 have understood me as intending in Sedley 1991), but that the goal of godlikeness is most fully achieved by the main cosmic structures, from the celestial spheres downwards, and only secondarily and derivatively by organisms like ourselves.

[40] Indeed, the one notable exception, widely cited, is *PA* 4.13.696b25–32, on the inconvenient positioning of the shark's mouth underneath, partly in order to spare other species.

little indeed to say about the global perspective. Like Aristotle's biological works, they are squarely focused on individual bodily functioning. Provided that Aristotle's physical and theological writings supply the missing global perspective on teleology, the relative restraint shown in the biological treatises should be read as symptomatic, not of his emancipation from Plato but, if anything, of a continuing debt to him.[41]

There is, as I have said, a pronounced reluctance among scholars to attribute to Aristotle any such global teleology, but various further considerations confirm that he intended it. In addition to the above passage (which as far as I know has not previously been invoked in the debate), and the matching passage from *Physics* 2.6 (195a5–13), in what follows I shall be briefly revisiting some of the other texts that point to a global teleology. Opponents of an Aristotelian global teleology have had no trouble in picking these passages off one by one, explaining each either as a temporary aberration on Aristotle's part, or as admitting of a more innocuous interpretation.[42] My reply is not that they cannot, at a stretch, be read in these alternative ways, but that their cumulative effect is far more powerful than that of any one taken individually, and such as to make any such escape implausible.

To this cumulative evidence, add the remarkable lack of counter-evidence. Precisely one passage has been cited as actually contradicting any idea of global teleology. This is *Ph.* 2.7.198b8–9, where in a summary of the four causes the final cause is summed up as 'because it is better thus, not absolutely (*haplōs*), but relatively to the being of each thing'.[43] If 'each thing' means each individual organism taken severally, Aristotle could indeed be denying a global teleology, even if he will at the same time unfortunately be rejecting his own thesis (*GA* 2.1.731b24–732a1) that animals' drive to reproduce is for the sake of perpetuating their species. But where does his emphasis actually lie? The terminology indicates that he is working with the familiar 'absolute–relative' distinction. So when he here asserts that the good served in his teleology is always relative, not absolute, his target could well be Plato's insistence that some decisions, for example

41 Cf. the valuable remarks in Falcon 2005, 8–9, about the probable dependence of Aristotle's order of exposition on Plato's.

42 I have defended a global-teleology interpretation in Sedley 1991 and 2000b, and I consider Cooper 1982, Kahn 1985, Code 1997, 130, and Matthen 2001 at least broadly in the same camp. The primary critics to whom I am referring here, are Wardy 1993, Judson 2005, Bodnár 2005, Johnson 2005; cf. also Pellegrin 2002, although he does not discuss the passages I rely on (he also seems to be incorrect, in n. 17, to place Sorabji 1980, 147 n. 8, in the former camp).

43 διότι βέλτιον οὕτως, οὐχ ἁπλῶς, ἀλλὰ τὸ πρὸς τὴν ἑκάστου οὐσίαν. The interpretation I offer here is largely identical to that in Sedley 1991, 190.

that the world should be spherical, were made by the Demiurge on the grounds that certain properties are intrinsically better than their alternatives, not because they do anyone or anything any good. Alternatively, or in addition, he could be extending from ethics to physics his critique of Plato's transcendent and absolute Good, whose alleged unattainability by us in his view makes it teleologically irrelevant (*EN* 1.6.1096b32–5). Either way his emphasis is on denying that his teleology involves any non-relative good. Hence there is no reason why, when he insists instead that the good served is 'relative to the being of each thing', his reference should be assumed to be limited to each discrete individual taken severally, and not more broadly to 'everything' – individuals, species and the world's entire contents alike.[44] This one passage then, in the absence of supporting evidence, falls far short of undermining the considerable weight of testimony in favour of a cosmic teleology, to which I now return.

In a much-discussed passage of *Physics* 2.8,[45] Aristotle maintains explicitly that in nature *all* regular events must be for the sake of something, and he includes under that generalization the specific case of rainfall. Summer rain, he allows, is rare, and he is ready to treat its occurrence as a mere accident. But winter rain, on which the growth of crops depends, is regular, and this he makes clear does serve a purpose. What purpose?[46] Elsewhere, when considering rainfall in its own right,[47] he is interested only in invoking the material processes of evaporation, condensation, and so forth. But that localized non-teleological perspective is perfectly compatible with another which sees weather as part of an inherently purposive cosmic nature, one

[44] This is, among other things, an answer to the question posed by Judson 2005, 360, why on an interpretation such as mine Aristotle would have written 'of each thing' rather than 'of something'. There is no reason to take the former to refer purely internally to individuals' self-benefit.

[45] I have discussed this passage at length in Sedley 1991, following the lead of Furley 1985.

[46] For the interpretation that the purpose served is water's return to its natural place, see Wardy 1993, 20–1. His primary text for this, *Cael.* 4.3.310a34–b16, may indicate no more than that the return of a simple body to its natural place is a return to 'form' in the literal sense of 'shape', and it is therefore probably not associating form with the final cause; but the chapter taken as a whole does, I think, tend to favour a teleological interpretation (e.g. the comparison to the potentially healthy becoming actually healthy, 310b16–19). Admittedly (cf. *GC* 2.10.337a1–15), the result of *all* elements returning to their natural places would be an entirely inactive and thus ungodlike sublunary world, but Aristotle presumably does regard it as better that at any rate most earth is at the centre, most water at its surface, most air above that, and most fire at the periphery. Since (cf. *Ph.* 8.2.252b21–3, 8.4.255a10–15, b29–31) water is not a self-mover, it must, strictly speaking, be moved by one or more external movers. Hence water's regular redistribution by rainfall to its proper place, the surface of the earth, will have as its moving cause the movements in the heavens (cf. *Mete.* 1.2.339a30–1, etc.). This makes rain not an autonomous mover driven by its own ends but part of an overall good cosmic distribution of the simple bodies (to the right places, at the right times, in the right quantities, etc.) orchestrated from the top down, and having among its innumerable beneficial outcomes the nourishment of plants.

[47] *Somn. Vig.* 3.457b31–458a1; *APo.* 2.12.96a2–6; *PA* 2.3.653a2–8; *Mete.* 1.9.346b21–36.

which among other things supports agriculture, as the reference to making the crops grow strongly suggests. We might think of Aristotelian rain as like human sweat.[48] Viewed in isolation, sweat is a liquid moved by purely material causes such as heating, cooling, evaporation and weight. Nevertheless, its occurrence is at the same time part of a purposive biological structure within which it serves an obviously beneficial end. In Aristotle's world, rain is very much like that.

A similar dual-perspective explanation accounts for another feature of cosmic teleology, one which Aristotle explicitly advertises in *Politics* 1.8, to the incredulity of many of his admirers. Lower species exist for the sake of higher ones (1256b10–22):

> Even at the moment of childbirth, some animals generate at the same time sufficient nutriment to last until the offspring can supply itself – for example all the animals which produce larvae or lay eggs. And those which bear live young have nutriment within themselves for their offspring for a time, the substance called milk. Hence it is equally clear that we should also suppose that, after birth, plants exist for the sake of animals, and the other animals for the sake of mankind – domesticated animals for both usefulness and food, and most if not all wild animals for food and other assistance, as a source of clothing and other utilities. If, then, nature makes nothing incomplete or pointless, it must be that nature has made them all for the sake of mankind (ἀναγκαῖον τῶν ἀνθρώπων ἕνεκεν αὐτὰ πάντα πεποιηκέναι τὴν φύσιν).

Here Aristotle follows Xenophon's Socrates (*Mem.* 4.3) in bringing the entire natural hierarchy within the scope of his teleology, which is itself given a manifestly anthropocentric focus. The food chain, and all other cases of inter-species dependence, are cases of nature's purposiveness, with man the ultimate beneficiary standing at the very top of the hierarchy. Once more, the 'nature' in question can hardly be identified with the natures of the individual plants and animals, or, for that matter, human nature. For Aristotle certainly does not think it is any part of the nature of the plants and lower animals to serve the interests of their predators, human or other;[49] and although it *is* part of human nature to exploit them, Aristotle's point is evidently not that here: for example, plants exist for the

[48] I choose this analogy as easier to illustrate than Aristotle's own chosen example, breathing *PA* 1.1.642a31–b4.

[49] On this cf. Pellegrin 2002, 312, who cites the case of the camel, pointing out that at *PA* 2.14.674b2–4 its tongue is said to be adapted to dealing with spiny plants, rather than the plants' being adapted to being eaten by the camel. But I see no conflict with the global-teleology interpretation, which is about how cosmic nature integrates the natures of individual divine and sublunary substances into a single system, and not about what determines those individual natures in the first place (see esp. Sedley 1991, 190–1).

sake of animals in general, he is telling us, and that aspect of the hierarchy could hardly be part of human nature. Rather it is the complex cosmic nature that is manifested in the world's inter-species ecology.[50]

The question has been asked[51] how this food-chain teleology can be reconciled with the internal teleology that dominates Aristotle's biological works. If pigs grow and function for slaughter and human consumption, how can they at the same time grow and function for their own wellbeing? This should not ultimately prove to be a problem for Aristotle. The natural strivings of living things are for survival, maturation and propagation. It is not in the pig's nature to queue up at the slaughterhouse. Nevertheless, the fact that the world actually contains pigs, and potentially contains roast pork, is a systematically beneficial feature of it, which Aristotle could not consistently with his teleological outlook consider merely fortuitous. Just as the nature of an animal can be invoked to explain why it has the parts that it does, so too the nature of the world, including its goal-directed structure with man at its apex, can be invoked to explain why it contains the species, weather systems and other amenities that it does. This in no way conflicts with strivings of a quite different kind, those of the pig itself, to perfect and perpetuate its own form rather than sacrifice those aspirations at the human dinner table. Here once more Aristotle's teleology can be best understood by adopting a dual perspective, combining the local and the global levels of explanation.

How, finally, can the world as a whole have a 'nature'? The natures in which Aristotle's chief interest lies are those of individual organisms, and the world is not for him, as it had been for Plato, a living organism. In *Metaphysics* Λ.10, however, when speaking of the world's nature as embodying cosmic good, he compares it not to the nature of an animal, but to the hierarchical structure of an army or household. And in *Politics* 1.2 he makes it clear that a city or household – itself a non-organic structure, despite consisting primarily of organisms – does have a nature, one which is indeed prior to the natures of its individual human components, since

[50] Judson 2005, 357–8 offers a new strategy for disarming the passage. Aristotle's words here are 'reflecting the viewpoint of the household manager or statesman: he is simply considering how many animals *can* be used for human ends, and reacting to the thought that some animals are, for example, too fierce to be so used'. But this does not sit very comfortably with Aristotle's insistence (however questionable) that nearly all animals are for the benefit of mankind, with no apparent emphasis on the rare exceptions. The last clause of the passage 'it must be that nature has made them all for the sake of mankind', also remains, in my view, a severe obstacle both to this and to two other contemporaneous attempts (Bodnár 2005, 23–4; Johnson 2005, 229–37) to read the passage as describing little more than how we do in fact adapt animals to our own use.

[51] Esp. Wardy 1993.

the latter are its parts. Specifically (1252b30–4),[52] this nature is the social or political system's completed form, also identifiable with its final cause. It seems then that any natural collective system composed of discrete natural substances, be it an army, a household, a city or a world, has as its 'nature' its own complex functionality, this being, irreducibly, an end over and above the individual functionality of its various components.[53]

52 διὸ πᾶσα πόλις φύσει ἔστιν, εἴπερ καὶ αἱ πρῶται κοινωνίαι. τέλος γὰρ αὕτη ἐκείνων, ἡ δέ φύσις τέλος ἐστίν. οἷον γὰρ ἕκαστόν ἐστι τῆς γενέσεως τελεσθείσης, ταύτην φαμὲν τὴν φύσιν εἶναι ἑκάστου, ὥσπερ ἀνθρώπου ἵππου οἰκίας.

53 For the close analogy between the hierarchical structure of such 'systems' and that of an individual organism like man, cf. *EN* 9.8.1168b31–2, 'Just as a city is thought to be above all its most authoritative element, and likewise every other system, so too in the case of man' (ὥσπερ δὲ καὶ πόλις τὸ κυριώτατον μάλιστ' εἶναι δοκεῖ καὶ πᾶν ἄλλο σύστημα, οὕτω καὶ ἄνθρωπος). For the notion of 'nature' in the *Politics*, including the analogy between political and zoological taxonomy, see also Lloyd 1993.

CHAPTER 2

Biology and metaphysics in Aristotle

Robert Bolton

THE AUTONOMY OF THE SCIENCES IN ARISTOTLE

Biology, for Aristotle, is an autonomous theoretical science. Or rather, to be more precise, biology is a proper part of physics or natural science (*phusikē*), which is itself wholly autonomous and separate from the other theoretical sciences. This means not only that biology is autonomous and separate from each of the mathematical sciences such as geometry or astronomy or optics. It is also autonomous and separate from the *primary* theoretical science, namely theology. Since, for Aristotle, the study of what is *qua* being, or general metaphysics, is either identical with or belongs exclusively to the same science as theology, it follows that biology is separate from and autonomous with respect to metaphysics in general. Since this relation of autonomy and separateness is symmetrical, if biology is separate from and autonomous with respect to metaphysics, for Aristotle, then metaphysics is equally autonomous and separate from biology.

Now one might think that these things are, or should be, quite uncontroversial, and hardly in need of saying, since Aristotle sets them out very plainly himself both in the *Posterior Analytics* and in *Metaphysics* Γ and E. According to *APo.* 1.10 there are just three types of ingredients that make up each science. As Aristotle says:

> Each demonstrative science restricts its concern to three things: (1) the entities whose existence it posits (*tithetai*) which fix the *kind* (*genos*) whose *proper attributes* it is its job to study; also (2) the so-called *common axioms* which are a primary basis for demonstrations; and thirdly the [proper] *attributes* [of the *genos*], of which it grasps what each signifies. (76b11–16)

Of these three types of ingredients which exhaust a given science, two – namely the entities that fix the special *genos* which determines the domain

It is a great pleasure to dedicate this chapter to my friend and colleague Allan Gotthelf, whose efforts have long led the way on these and related matters.

of the science, and the proper attributes of that *genos* – are, Aristotle also says in 1.10, entirely unique to that science and shared by no other science. These items are introduced either in definitions or in posits or hypotheses both of which are always proper to a particular science (76a37ff.; cf. 1.2.72a14–24). So the only content shared by different sciences consists of the remaining ingredient, namely the so-called common axioms such as the principle of non-contradiction. That is why these axioms are called common by contrast with the other ingredients which are special. So *nothing* beyond these common axioms figures, or can figure, in more than one science. There is no room at all there for the substantive material of metaphysics or of biology to figure in the other science.[1]

In *Metaphysics* Γ and E Aristotle offers the same basic account of the autonomy of the sciences as in the *Analytics*. He says again, at the beginning of E.1, that each of the so-called special sciences, such as the individual mathematical sciences and natural science, marks off one special genus or part of what is and studies only this genus and its proper attributes (1025b3ff.). He does also, in E.1 and in Γ.1, assign to metaphysics a certain generality which distinguishes it from these other sciences on the ground that, in a way, metaphysics studies all the things that there are. But, as is well known, he insists that metaphysics only studies these things that are in a special restricted way, namely *qua* being (1003a21–32). That is, as he indicates in Γ.2, metaphysics studies the things that are which happen to be substances *qua* substances; and it studies the things that are which happen to be qualities *qua* qualities of substances; and so on (1003b5–16; cf. Z.1.1028a35–6; Z.4.1030a17ff.). The general subject matter of metaphysics is strictly restricted to this and, importantly, Aristotle also insists in E.1 that none of the other sciences is at all concerned with this (1025b9–10). So, in sum, general metaphysics is only concerned with the study of what is *qua* being and the proper attributes of what is *qua* being, and none of the other sciences is concerned with this. Some natural scientists, Aristotle notes in Γ.3, did stray from these strictures and undertook to study general truths that properly apply to what is *qua* being, such as the principle of non-contradiction, as though they were a part of the subject matter of physics (1005a31ff.). But, he argues, they were in error and, he indicates,

1 There are, according to Aristotle, sciences that are subordinate to others, as for instance optics is to geometry, where material from one higher science may figure in its subordinate because the two, in part, share a common genus. See *APo.* 1.13.78b35ff.; 1.7.75b8–9. But Aristotle never posits any such relation between metaphysics and biology, or vice versa, and, as we shall immediately see, such a relation is excluded by what we find in the *Metaphysics*. So we may ignore this exception here. Cf. Bolton 1995, sec. II. The present discussion is intended to extend and further articulate certain of the proposals developed there.

such errors were "due to a lack of education in the *Analytics*" (1005b3–4). So despite the special sort of generality that does characterize metaphysics by contrast with geometry or natural science it is clear here again, just as in the *Analytics*, that there is no room at all for overlap in content or subject matter, for Aristotle, between metaphysics and biology or physics.[2]

However, Aristotle's seemingly plain words on these points have not prevented many recent writers from finding in his biology, or in his natural science more generally, crucial contributions to his metaphysics in the strict sense, even contributions which correct doctrines in the *Metaphysics* itself. Some have argued, for instance, contrary to what we find in *Metaph.* Z.16 (1040b5ff.), that we learn directly from the biological works that the substances, in the metaphysical sense, for Aristotle, are, or at least include, the *parts* of animals such as hands and feet, flesh and blood. Others have argued that what we rather learn from the biological works, contrary to what we find in, for instance, *Metaph.* Z.7 and 8 (1032a18–19; 1034a4), is that the unqualified substances or primary realities are not particular plants and animals *or* their parts but rather are certain particular biological *forms* such as those that are transmitted from father to offspring which account for various special inherited family resemblances (see *GA* 4.3). Still others have argued that what we learn from the biological works is that Aristotle was in the end deeply ambivalent on the question of whether the unqualified substances are these particular forms, or rather, instead, certain species forms.[3] More generally, many recent writers have held that certain key doctrines from Aristotle's biology or physics, such

[2] Even the principle of non-contradiction, as a truth of metaphysics concerning everything that is *qua* being, is not supplied by metaphysics to physics since physics itself posits this principle, Aristotle says, as and only as covering its own *special* genus, and in that specific form the principle is not the concern of or validated by metaphysics (*Metaph.* Γ.3.1005a25–7; cf. *APo.* 1.10.76a37–b2). It has often been argued that Aristotle's requirements in the *Analytics* for a science, including those that there guarantee the autonomy of the different sciences, are abandoned in the *Metaphysics*. In particular, it has been claimed that the doctrine of the *Analytics* that each science deals with a unique genus (and its proper attributes) is rejected in the *Metaphysics*. Whether or not this is so, however, it is clear that in the *Metaphysics* itself, for the reasons just given, Aristotle holds that, as the study of what is *qua* being, the subject matter of metaphysics does not overlap with that of natural science. It is also arguable that Aristotle there in fact accepts and even argues for the view that like the other sciences metaphysics too deals with its own special genus, so that in this respect metaphysics does not at all depart from this requirement for an autonomous science in the *Analytics*. This special genus for metaphysics is not, of course, being, since, for Aristotle, being is not a genus, but is, rather, substance. For defense of this view with discussion of the more common alternative see Bolton 1995 and Code 1996.

[3] For recent presentation and discussion of these various views see the contributions of D. Balme, J. Cooper, M. Furth, G. Lloyd and P. Pellegrin to Gotthelf and Lennox 1987 and to Devereux and Pellegrin 1990, with further references there. See also, more recently, Reeve 2002, Sharples 2005b and Henry 2006a and 2006b, with further references there.

as his so-called hylomorphic analysis of the twofold nature (*phusis*) of natural objects, are directly taken over by him from his biology and physics into his metaphysics. These doctrines from physics, so it is claimed, then determine, to a significant degree, the central teachings of *Metaphysics* Z–H.[4] We can see this, it is sometimes suggested, in *Metaph.* Z.7–9 where Aristotle introduces aspects of a *phusikos* discussion, proper to natural science, in order to reach crucial aspects of his metaphysical doctrine. On one standard reading, this new material from biology or physics requires Aristotle to abandon the view of the *Categories* that particular plants and animals are primary substances and also the doctrine that the species and genera of these plants and animals are secondary substances.[5]

In addition to such supposed direct infusions from biology or physics into metaphysics, it has also been argued, from the other side, that Aristotle presents and defends in his *Metaphysics* certain crucial biological theses, even biological theses which supersede those in the biology or physics itself. For instance, many have claimed that, contrary to what we find in the physical and biological treatises (e.g. in *Ph.* 2.2.194a1–7; *de An.* 1.1.403a25–b9; *PA* 1.1.640b24–9, 642a14–31), we learn from *Metaphysics* Z in particular that there is no place in the essence or strict definition of a biological human being as such for a reference to its matter or material cause.[6] Rather, it is argued, we can see from Z.10–11, and from Z.17 in particular, that it is the form alone of a human being that is causally responsible, in the strict sense, for its belonging to the human species, for its *being* as a biological human being. More generally, Z.17 shows us, so it is claimed, that for Aristotle it is the form alone of any natural kind that is ultimately *what it is* as that kind of thing.[7] Still others have argued that we can see from, for instance, *Metaph.* H.6, again contrary to the physical and biological works (e.g. *de An.* 1.1.403a25–b9), that the form of a human being includes or is identical with its matter or at least that the form and matter of a human being have the same essence.[8] Still others have argued that we can see from *Metaph.* Θ.7 that Aristotle has revised his requirement in the *Physics* for the persistence of matter in all natural change.[9] Some have even claimed that we learn from the *Metaphysics* that for Aristotle it is necessary to understand the nature of god, that is, to master theology,

[4] Such an approach is developed, e.g., in Burnyeat 2001. [5] See, e.g., Lewis 1991.

[6] See, e.g., M. Frede in Devereux and Pellegrin 1990 and Frede and Patzig 1988.

[7] See, e.g., Lewis 1991, ch. 7 and Charles 2000.

[8] See, e.g., the contributions of D. Balme and A. Kosman to Gotthelf and Lennox 1987. See also the essay by M. Gill in this volume.

[9] See, e.g., Gill 1989.

in order to understand scientifically what it is to be a human being or a member of any other biological or natural kind.[10]

Serious questions for all such claims are raised immediately by what we find in *APo.* 1.10, and in *Metaph.* Γ.1–3 and E.1. Given what Aristotle says there about the autonomy of biology *and also of metaphysics*, it cannot be the case that the results of metaphysics, strictly speaking, place *any* essential constraints at all on the biologist or natural scientist as such, nor that the results of biology or natural science, strictly speaking, place *any* essential constraints at all on the metaphysician as such. Rather, as these texts show, apart from the common axioms, each domain has its own unique data for explanation and its own unique principles for the explanation of its unique data. Each of these – both data and principles – may be learned, and must be mastered, in a way that is completely independent of knowledge of the other domain and its different data and principles.

This does not mean, of course, that metaphysics and biology are permitted to contradict each other. Aristotle's views on meaning and truth clearly rule that out, as we know well enough from, for instance, *Metaph.* Γ.4. Aristotle undoubtedly intends all of his views to be consistent. So on this basis what we find in the *Metaphysics* may and should constrain *our interpretation* of what we find in the biological works, and vice versa. Furthermore, we should expect that since Aristotle's own views in any one area, by virtue of his commitment to logical consistency, will constrain his own thinking in other areas, indications of this will from time to time appear in the *Corpus*. Nothing prevents Aristotle from relying on or drawing attention to the fact that his results in one area cohere well in various ways with his results in another, for legitimate heuristic or persuasive purposes. So *Metaph.* Γ.1 and E.1 do not rule out that we may find various metaphysical remarks in the biological works, or biological facts, as Aristotle sees them, stated in the *Metaphysics*.

There are also two more important ways in which, for Aristotle, material from outside the strict domain of a given science may at least guide and influence inquiry in that science. As we have noted, Aristotle indicates explicitly in *Metaph.* Γ.3 that his procedures there are guided by his own doctrines on the nature of science and of scientific inquiry in the *Analytics* (1005b2–5). He applies this straight away in Γ.3 when he argues, as in the *Analytics*, that such common axioms as the principle of non-contradiction cannot be strictly demonstrated in any science that employs them (cf. *APo.* 1.3). We find a similar pattern of thought in *PA* 1.1 where Aristotle offers

[10] See, e.g., M. Frede in Romeyer Dherbey 1996.

various guidelines on the proper way to pursue inquiry in biology and natural science. Some of his methodological rules there are common to all sciences and repeat material from the *Analytics*, such as the rule that in natural science as even in mathematics one should first inquire into the *phainomena* presented by one's subject matter and then go on to give the explanations and causes of those *phainomena* (639b7ff. with 640a13–15; cf. *APr.* 1.30.46a17–22). That is, we may say, such rules as this in *PA.* 1.1 belong to general analytics or general philosophy of science. Other rules of method, however, as Aristotle also indicates in *PA.* 1.1, are more special to natural science, such as certain of those that concern material and final causes, form and matter, and actuality and potentiality, all of which figure prominently in *PA.* 1.1. But these rules too are assigned there by Aristotle not to natural science itself, or to metaphysics, but rather to a special *paideia*, or educated ability to judge correct methodological procedures, a competence which is different from the grasp of any science (639a1–18). So these more special rules also belong to analytics, what we may call special analytics or special philosophy of science. Thus, contrary to what is often claimed, these points of scientific method in *PA.* 1.1, even the ones that may be of special interest to natural science, such as those that concern form and matter, are not a *part of* natural science, or of any other science, even though Aristotle can and does rely on them in important ways to guide his inquiries in natural science. Equally, appropriate rules from general or special analytics may guide Aristotle's inquiries in metaphysics. In *Metaph.* Γ.3 Aristotle assigns the grasp of such rules again to *paideia* not to any science (1005b3–4).[11]

There is also one final way, worthy of special attention, in which remarks or facts from one autonomous domain may easily influence the investigation of another for Aristotle. In his scientific inquiries, both biological and metaphysical, Aristotle often introduces or pursues a topic by means of an exploratory aporetic discussion. In such discussions he sometimes, at least, uses his dialectical method of inquiry. Dialectic, however, has no restrictions at all on its proper subject matter, as Aristotle makes clear in, for instance, *Sophistical Refutations* 11 (172a12ff.), and also in *Metaph.* Γ.2 (1004b17ff.). In this way dialectic is quite different from science, and dialectical reasoning and inquiry are quite different from strict scientific reasoning and inquiry. *Any premises from any domain at all* may be used in *dialectical* reasoning *on any topic at all*, providing only that these premises are *endoxa*, or noted and accredited opinions, of an appropriate sort (see

[11] On this topic, see also Lennox in this volume (ch. 3). Contrast Burnyeat 2001.

Top. 1.1, 8.1–6; *APo.* 1.11.77a31ff.). This means that in any dialectical passages, either in the biological works or in the *Metaphysics*, we may well find endoxical material from the other domain. After all, there are commonly accepted facts or *endoxa* which belong to biology, such as "humans generate humans," and "humans have two feet," and "humans have a certain goal or end." There are also commonly accepted facts or *endoxa* which belong to metaphysics, according to Aristotle at any rate, such as "plants and animals are independent realities, namely substances, and colors are not," and "the predicables which fall into different categories indicate either what something is, or how much, or what sort, or where, or when, etc." (see *Metaph.* Z.2.1028b8ff.; Z.8.1033b29ff.; H.1.1042a3ff.; *Top.* 1.9). But this endoxical material cannot be introduced in any dialectical passages as material with scientific standing, either with the scientific standing it may have in its own proper discipline or, perhaps more importantly, with any scientific standing in the discipline at hand. Aristotle contrasts results based on dialectical reasoning "in accord with received opinion" with scientific results reached by reasoning "in accord with truth" (*APo.* 1.19.81b18ff.; cf. *Metaph.* Γ.2.1004b25–6).

What this means is that when or if alien material is used dialectically in inquiry in a discipline in which that material has not been given scientific standing, no results based on this material can as such themselves have proper scientific standing in the context in which they are used. Generally speaking, they can only have heuristic value for strict scientific inquiry by serving or tending to rule out certain options or to suggest relevant data or promising approaches (see *Top.* 1.2.101a34ff.). The direct implication of this is that we cannot learn any truths of biology *as such*, that is, as scientific truths of biology, from Aristotle's *Metaphysics*, either from any dialectical passages there, where such truths cannot have scientific standing, or from any strictly scientific metaphysical passages in the *Metaphysics* which cannot concern biological matters. This means that none of the strictly scientific explanatory work of biology can essentially depend on or be done in Aristotle's metaphysical inquiries, either in the dialectical or in the non-dialectical parts of those inquiries. Equally, none of the strictly scientific work of metaphysics can essentially depend on or be done in Aristotle's biological inquiries. In this sense, for Aristotle, metaphysics and biology, or natural science, place *no constraints* on each other. Nevertheless, as we have noted, very many interpreters of Aristotle have ascribed to him a different view. So in this discussion I would like to explore some of the more important claims already mentioned according to which Aristotle would appear to violate his strictures concerning the autonomy both of metaphysics and biology. This will, I think, also help us to make progress on various questions of special importance both for our understanding of Aristotle's biology and also of his metaphysics.

CAUSAL EXPLANATION IN *METAPHYSICS* Z.17

Let us consider first then the claim made by a number of writers that one of Aristotle's main aims in *Metaphysics* Z–H is to give us an analysis or a strict definition of what it is to be a biological human being and, more generally, of what it is *to be* for any natural kind of thing as such. Aristotle proposes to give us in Z–H, as he himself says, an analysis of what it is to be a perceptible substance (Z.3.1029a33ff.; Z.11.1037a10ff.). A human being is a paradigm of such a substance and thus, it is often supposed, Aristotle focuses, for instance, on this paradigm and tells us what it is, strictly speaking, that makes something a member of the human species. His answer is particularly apparent in Z.17, so it is often argued, where Aristotle shows us that it is human *form* alone that is the essence of and is causally responsible for the fact that Socrates, for instance, is a biological human being. More explicitly, as some say, it is the presence of Socrates' form to his matter that causes him to be a member of the human species. Aristotle then generalizes from this case, and others like it, in Z.17, so it is claimed, and argues that if form is what causes Socrates who is a perceptible substance, to be a member of his biological kind, the human species, and similarly if form is what causes other perceptible substances in other biological kinds to be members of their species, then it is in general form that is substance.[12] That is, Aristotle here draws a certain metaphysical conclusion, that substance is form, from certain biological data.

This currently common reading of the argument of Z.17 follows the earlier lead of Ross, who claims:

> The general line of thought in the chapter [Z.17] is as follows: Substance is a cause. Therefore, if we can find out what in general is the cause of things, the [general] answer to the question "Why?," we shall have found what substance is . . . In general it [the answer to this question "Why?"] is a statement of form or essence. Form or essence, then, is substance.[13]

So framed, Ross's account of the argument of Z.17 faces various difficulties. To begin with, on Ross's construal Aristotle's argument is formally invalid. Even if we take it as a premise that in general the ultimate answer to the

[12] See, e.g., Charles 2000. [13] Ross 1924, II.224.

question "Why?" for any kind of thing, is "a statement of form," this must mean that the ultimate answer to the question "Why?" for any kind of thing, is a statement of *its* form or *its* formal cause. From this it follows not that the answer to the question "Why?" in the case of substance, is form *simpliciter*, as Ross supposes, but only that the answer will give the form or formal cause *of substance*, that is, of the kind substance, whatever that formal cause may be. We will return to the question of the validity of Aristotle's argument later. More to the point for present purposes, in any case the main premise of the argument, on Ross's account, is in fact rejected by Aristotle in Z.17 itself. As we have just noted, this main premise is that the ultimate answer to the question "Why?" concerning any type of entity at all, will give the essence in the sense of the form or formal cause of that entity. But this is certainly not Aristotle's view. In Z.17 he explicitly says that in the case of generated things the ultimate answer to the question "Why?" which gives the essence or ultimate cause of the thing, will give the efficient cause of the thing, not the form or formal cause (1041a27–32). The example of thunder, mentioned in Aristotle's discussion, is a clear case in point since the ultimate cause of thunder, namely the quenching of fire in the clouds, is its efficient cause (1041a24–6 with *APo.* 2.8.93b7–12). Elsewhere, in *Ph.* 2.7 and *APo.* 2.11, Aristotle argues more generally that instances of *each* of the four causes may serve as the ultimate cause which answers the question "Why?" in certain cases. So the ultimate cause and essence of a thing for Aristotle is not always its form or formal cause, either in Z.17 or elsewhere. However, we may avoid this difficulty and still follow Ross's basic approach, as many more recent writers do, if the generalization which leads to the metaphysical conclusion that substance is form is based only on the answer to the question "Why?" in the case of the various species of perceptible substances, such as human beings and horses (and perhaps also of artifacts such as houses, as some suppose; both of these types of cases, in addition to thunder, are prominently discussed in Z.17). Clearly, however, if this *is* Aristotle's line of argument then, as we have seen, his metaphysical conclusion, that substance is form, will still be crucially and essentially based on certain biological facts concerning why the various types of living things are members of their biological species. This would definitely violate Aristotle's strictures on the autonomy of metaphysics according to which metaphysical conclusions cannot be essentially based on or inferred from biological data.

In considering further the merits of this standard reading of Z.17 it is crucial to keep in mind that Aristotle's own question there, as in Z as a whole, is "What is substance?" As he tells us at the very beginning of the chapter, he understands this question about substance, as to what it is, as a causal question, that is as the question "What causes something to be a substance?" (1041a6ff.). As is often pointed out, by Ross himself among many others, in Z.17 Aristotle clearly introduces and purports to follow the general procedures for reaching a scientifically adequate answer to any question of the form "What is X?" that are laid out in detail in book 2 of the *Posterior Analytics* where, as in Z.17, this type of question, "What is X?," is explicitly understood as a causal question (see *APo.* 2.2). This connection with *Posterior Analytics* 2 gives us, in fact, our best clue as to why Aristotle thought it important to write Z.17. This is something of a puzzle. The summary of the contents of Z that is offered at the beginning of H makes no reference to Z.17. Moreover, the question that Aristotle raises anew in the chapter, namely, "What is substance?," is a question that he seems to have already answered. He says himself at the beginning of Z.17:

> *What* it is necessary to say substance *is*, and *what sort* of thing, let us say again, making in a way a fresh start. (1041a 6–7)

Here Aristotle indicates that he has already said what substance is, for instance, presumably, in Z.10 and 11, where he says, in no uncertain terms, that substance is form (1035b14–16; 1037a21–30).[14] So why does Aristotle feel it necessary to repeat this in Z.17? One main reason, at least, seems to be that in *Posterior Analytics* 2 Aristotle provides us with his general doctrine concerning the right way, indeed the only right way, to reach an answer to a scientific question of the form "What is X?" Whatever progress he may have made earlier in Z, he has not made it clear there just how his answer to the question "What is substance?" fits the necessary requirements for a proper answer to that question on which he strongly insists in *Posterior Analytics* 2. In Z.17 he shows us how. In brief, according to the requirements of *Posterior Analytics* 2 one reaches a final answer to a definitional question of the form "What is X? (*ti esti;*)," when one produces a special kind of definition, the kind that is revealed in what Aristotle calls, in *APo.* 2.10, "a syllogism of what something is (*sullogismos tou ti esti*)" (94a12, cf. a2). In such a syllogism the major term, namely the predicate term of the conclusion, designates the entity to be defined, and the middle term designates the ultimate cause of this entity, which ultimate cause is, in the most fundamental sense, *what* this entity *is*, what its essence is (see

[14] I take *palin* at 1041a6 with *legomen*, contrary to Ross and others. Cf. *APo.* 2 19.100a14–15: "Let us say again what was said earlier, but not said clearly." Some would argue that the earlier chapters, Z.10–11, are aporetic and inconclusive. This approach would not affect the account of Z.17 offered here, nor would the alternative approach offered in Burnyeat 2001.

APo. 2.1–2, 8–10, 16–17). We know from Z.17 what the ultimate cause is which provides the final answer to Aristotle's definitional question "What is substance?" As he says:

Our inquiry [in asking "What is substance?"] is for the cause (*aition*) of the matter, namely the form, by virtue of which [cause] it [the matter] is (a) something (*ti*). This [i.e. form] is [what] substance [is]. (1041b7–9)[15]

If *form*, then, is the ultimate cause and thus the middle term in the required syllogism that reveals *what substance is*, what are the other two terms of the syllogism, the terms that make up the conclusion of the syllogism? Or, in plainer language, what is the fact to be explained of which form, as middle term, provides the ultimate cause? In the text this fact to be explained is clearly identified as the fact that "the matter is (a) something (*ti*)" (1041b8). What does Aristotle mean by this? As we have noted, many have supposed, following Ross, that Aristotle means here that the fact to be explained is, for instance, the fact that Socrates is a human being, or, more specifically, as some say, the fact that Socrates' type of body has such characteristic human features as, say, two-footedness, and other *biological* facts of that type. If this is correct then Aristotle's initial "syllogism of what (a) substance is" would go roughly as follows:

The human body or matter has human form.
What has human form is two-footed, etc.
So, the human body or matter is two-footed, etc.

From this inference, which is presumed to show us that human form is what is responsible for the characteristic features of the human species,

[15] This translates the text of the MSS, which is followed by all the ancient commentators. The phrase "namely the form" (*touto d' esti to eidos*, 1041b8) is excised, first by Christ 1885 and, following him, by Jaeger 1957 and Frede and Patzig 1988 on the grounds that it interrupts the flow of thought in the text, makes the antecedent of the relative pronoun in b8 unclear, and opens the door to misunderstanding of Aristotle's meaning. Perhaps it does, but these faults of exposition, if such they be, are not uncommon in Aristotle, even in Z.17. Note the very similar interruption of thought at 1041a 28, *touto d'esti to ti en einai*, which Jaeger also excises, but Christ, Ross, and Frede and Patzig retain. This phrase also makes the antecedent of the relative pronoun there unclear. See also 1041b11–25 for other more abrupt interruptions of thought in the chapter. Since the phrase in question at b8 is in all the manuscripts and in all the ancient commentators, and it can be made very good sense of, following Ross's suggestion that it be treated as, in effect, an appositive, or as parenthetical, it is surely preferable on philological grounds to keep it. More strongly, if the phrase is excised then Aristotle does not do in the chapter what he earlier says he is going to do, namely to "say again *what* substance *is*." Nor does he produce the "syllogism of what substance is" which the discussion leading up to 1041b7–9 directly prepares us for. In any case, however, since even those who would excise the phrase agree that it gives us Aristotle's actual doctrine, we may reasonably ask how we are to interpret his doctrine so understood. Those who wish to excise the phrase may read the discussion here in this way.

together with other inferences like it concerning other species of perceptible substances, Aristotle is presumed to infer inductively that in general substance is form.

However, according to the views on scientific method in the *Analytics* that Aristotle clearly invokes here in Z.17, in order to properly reach the ultimate middle term which answers a scientific question of the form "What is X?," understood as a causal question, one must start from some fact to be explained in giving this answer which articulates the existence of Xs by specifying *proper features* of Xs *as Xs*. As Aristotle says in *APo.* 1.6 and 9:

> Scientific demonstrations are concerned with what holds of things in themselves and they proceed from such things . . . It is clear [then] that one cannot demonstrate something about a given thing except from the principles which concern that thing, since what is being established [to belong to it] must belong to it *as that thing* . . . We have non-accidental scientific [demonstrative] knowledge of something, then, when we know it by grasping that in virtue of which it holds *in respect of that thing*, from the principles which belong to that thing *as that thing*. (75a29–31, b37–76a6)

This doctrine is reaffirmed in *Metaph.* E.1 and Γ.2 in specific application to metaphysics:

> We are seeking the principles and the causes of the things that are but, clearly, of them *qua* being. (1025b3–4)

> For just as there are proper attributes (*idia pathē*) of number *qua* number . . . so also there are certain proper attributes of what is *qua* being and it is about these that the [first] philosopher investigates the truth. (1004b10–17)

> These [attributes whose causes are investigated by metaphysics] are *as such* attributes (*kath' hauta pathē*) of what is *qua* being and not of what is *qua* numbers or lines or fire. (1004b5–6)

Aristotle develops this doctrine more fully, in a general form, in *APo.* 2.8, where he argues:

> It is impossible to know [i.e. to answer the question] what something is if we do not know whether the thing exists . . . When, however, we only know accidentally that it exists, our grasp is necessarily in no way directed to what it is . . . But when we grasp *something of the thing itself* [and know that it exists by virtue of grasping that] our task is much facilitated. (93a19–28)

We then learn *what* something *is*, Aristotle indicates, when we find the middle term that explains the special fact of which we are aware when we are aware of the existence of the thing by grasping something of the thing itself (93a36–b14; cf. *APo*. 2.1–2). Aristotle goes on to make this more precise in *APo*. 2.10 by arguing that the fact to be explained in searching for the answer to the question "What is X?" itself gives us one kind of *definition* of X. That is, the *conclusion* of the "syllogism of what something is," Aristotle says, itself provides one kind of definition of that thing itself. For example, in the case of thunder we find the answer to the question, "What is thunder?," by finding the middle term which explains the fact, the conclusion, that a certain noise occurs in the clouds. Here, as Aristotle says, the fact to be explained, noise in the clouds, offers us one type of definition of thunder, one account of the thing thunder itself or of what it itself is (93a20–9; 93b38–94a9). This same doctrine reappears, very clearly, in Z.17, where Aristotle emphasizes again, at some length, that we will not succeed in answering the causal question "What is X?" for any X, unless we have properly articulated the right sort of fact to be explained in answering the question. As he says:

> The object of the inquiry [in asking "What is X?" for any X] easily escapes our notice in cases where [in our question] items are not predicated one of another, for instance where we ask "What is man?," because the question [in this case] is put [too] simply and not by defining (*diorizein*) [man by saying] that this is these things [and then asking "Why?"]. (1041a32–b2)

So here in Z.17 Aristotle emphasizes again, just as he does in *APo*. 2.8 and 10, that the features of a thing the connection between which is *explained* in finding its ultimate cause and essence are defining features of one sort of the thing itself, features which belong to it as that kind of thing, whether the kind in question be thunder, or the human species or substance. It is, as in the *Analytics*, the grasp of such a fact that gives us the sort of knowledge of the *existence* of our object of inquiry that permits us to discover its cause and essence (1041b4ff.). So, for instance, to illustrate this now in a metaphysical case, if we want to answer the question "What is a *quality*?," we must start from some fact or facts to be explained which articulate the existence of qualities by introducing specifying, or defining, features of qualities *as qualities*, and by proceeding to explain the connections among these features. According to *Categories* 8, for example, it is a distinctive feature of qualities as such that they are attributes in virtue of which things are called *like* and *unlike* (11a15–19). To answer the question "What is a quality?," then, one must explain why this sort of thing is so: why the various qualified things as such are able to be *like* and *unlike*. It would be quite inappropriate to proceed, in order to answer the question "What is a quality?," by first trying to answer the question "What is thunder?," even though thunder, as a certain sort of sound, *is* a quality for Aristotle. To answer the question "What is thunder?," for him, as we have just seen, we must start from proper knowledge of defining features of a sort of thunder as such, for example that it is a certain sort of noise and that it occurs in the clouds, and proceed to explain why these features are connected (*APo.* 2.8.93a20ff., b7ff.). In such a case we do not introduce any defining features proper to qualities as qualities, such as those we find in *Categories* 8. So we cannot answer the question "What is a quality?" by first answering the questions "What is thunder?," "What is aroma?," "What is conical shape?," and so on, and then somehow trying to generalize from our answers about these particular qualities since in this procedure we do not start from or introduce any specifying or defining features proper to qualities *as qualities*. Indeed, it is clearly not the business of metaphysics even to know what thunder is, or what aroma is, or what a cone is, so the metaphysician as such would be at a loss to know how to proceed in this way. Similarly then, for Aristotle, one cannot properly answer the question "What is substance?" or "What is a perceptible substance?" by first answering the questions "What is a human being?," "What is an octopus?," "What is a bee?," and so forth, and then somehow generalizing from this. To answer these latter questions one must explain why certain specifying or defining facts that characterize human beings, or octopuses, or bees as such are so, for instance the fact that the human body is two-footed, or that the octopus's body is eight-footed (as the Greeks say), or that the bee's body is four-winged. But, again, the metaphysician is in no position at all to explain these facts or even to know that they are the relevant facts to be explained in answering the questions "What is a human being?" or "What is an octopus?," or "What is a bee?" Such facts as these are the province of the biologist. Such potential conclusions of demonstrations as these, in Aristotle's framework, give us definitions, of a sort, of what a human being is or of what an octopus or a bee is, not of what a substance is. Indeed, the very suggestion that in order to learn what substance is the metaphysician must know how to explain scientifically why octopuses have eight legs (as we say) and why bees have four wings is more than a little absurd. Rather, to answer the question "What is a substance?," the metaphysician must explain why certain specifying or defining features that belong to substances as substances are connected. This is what Z.17, following the *Posterior Analytics*, tells us.

Moreover, even if it were to turn out, for Aristotle, that the only substances, or the only perceptible substances, *are* necessarily the members of the various biological species, such as the human species, it would not follow that what ultimately makes these things substances, namely independent self-subsistent realities, is the same thing as what ultimately makes them the members of their biological kinds, since what explains why things are members of one kind does not necessarily explain why they are members of another kind even if those kinds are necessarily co-extensive. For example, in *APo.* 1.13 Aristotle treats the two kinds *non-twinkling heavenly body* and *nearby heavenly body* as necessarily co-extensive kinds. But he clearly indicates there that the cause of the non-twinkling of these bodies, the planets, is not the same as the cause of their nearness. In fact, their nearness, Aristotle says, is the cause of their non-twinkling, but their nearness is certainly not the cause of their nearness. So Aristotle's presumed inference here, on Ross's type of account, from the supposed fact that their form is what *causes* the members of the various biological species to be members of their species to the conclusion that form is what *causes* all these living things to be substances is an invalid inference, by Aristotle's lights, even if the two kinds are necessarily co-extensive. Indeed, according to Aristotle's views on scientific method, the ultimate middle term that answers one scientific question of one science such as "What is substance?" must *uniquely* answer that question and cannot also answer another different scientific question from another science such as "What makes something a member of a biological kind?" or "What makes something a living thing?" The ultimate middle term that applies to one kind, for example substance, is what ultimately *defines* it, and two quite different kinds from different sciences cannot have the same ultimate *definition* for Aristotle without these two kinds being intentionally as well as extensionally the same, and that is impossible for him since it would violate his doctrine that the ultimate definitions of a science are unique to it. As Aristotle says in *Metaph.* Λ.5:

> The causes and elements of different things that are not in the same kind, such as colors, sounds, substances, and quantities, are different, except analogically [i.e. by playing the same sort of role in the different cases]. (1071a25–7)

So if form is *what* substance *is* – what makes something a substance – then form cannot also be *what* a biological kind or living thing *is*. (That the very *same* scientific explanation cannot apply to two distinct phenomena or answer two distinct questions is, of course, a plausible principle in itself, not simply a main principle of Aristotle's philosophy of science.)

So when Aristotle answers the question "What is substance?" in Z.17, by saying that it is the possession of form that explains primarily why a certain type of material thing, such as Socrates, or his type of matter or body, is "(a) something (*ti*)" he must mean that form explains not why Socrates, or his type of body, is a human being or has two feet, and so on, but that form explains why the relevant matter has some specifying or defining feature proper to all substances *as substances*. If this feature is not being a human being, or having two feet or any more general biological feature of that sort, such as having an organic body, what might it be? For an answer we are aided by a passage earlier in Z.7 where Aristotle says:

> Everything that comes to be . . . comes to be something (*ti*) and the something (*to ti*) which it comes to be, I say, may be found in any category. For either it [the something] is a *this* (*tode*), or a *how much*, or an *of what sort*, or a *where*. (1032a13–15)

This passage shows us that one of the things that Aristotle might well mean, and expect us to understand, when he says in Z.17 that a certain matter or body is "(a) something (*ti*)" (1041b8), is that it is a *this* (cf. Z.8.1033a24–32; Λ.2.1069b9–11). However, of course, being a *this*, as Aristotle often indicates, *is* a distinctive specifying feature of substances as such (*Cat.* 5.3b10ff.; *Metaph.* Z.3.1029a28; Z.4.1030a2–6). Moreover, we know directly from *de An.* 2.1 that for Aristotle the possession of form does explain why matter of the appropriate sort is a *this*. As he says:

> We say that substance (*ousia*) is one kind of thing, in one way as matter which is not in itself a *this*. In a different way, substance is shape and form which is that by virtue of which something [which possesses it] is straightaway said to be a *this*. In a third way, substance is the compound of these things. (412a6–9)

Here Aristotle tells what he means when he speaks of substance in the sense of form. He means that entity by possession of which some suitable subject is a *this*. These passages offer us, then, a reading of our text in *Metaph.* Z.17 that fits precisely those requirements of the *Analytics* that are there invoked. According to those requirements, as we have seen, to find the ultimate middle term that accounts for why something is a substance one must start from a very specific sort of knowledge, namely proper scientific knowledge *that there are substances* which we only have when we grasp some specifying or defining feature of substances that belongs to substances as substances (*APo.* 2.8). To know that Socrates, or his type of body or matter, is a self-subsistent *this* is to know scientifically *that there are substances*. To know that Socrates, or his type of body or matter, is a human being or a certain sort of animal, such as a two-footed animal, is to know scientifically, as at *APo.* 2.8.93a24, that there are human beings, but not at all to know scientifically that there are substances. So the former fact not the latter, is the right sort of fact starting from which one finds the answer to the question "What is substance?" according to Aristotle's views on scientific inquiry. The genuine "syllogism of what substance is," then, will go roughly as follows:

Socrates' type of matter or body has form.
What has form is a *this*
So, Socrates' type of matter or body is a *this*.[16]

The generalized definition of substance, or of perceptible substance, corresponding to this syllogism would then be this:

A perceptible substance is, by definition, what is a *this* due to the possession of *form* by its body or matter.[17]

PREDICATION AND CAUSATION IN THE *METAPHYSICS*

To see further how different these two questions are, namely "What makes something a substance?" and "What makes something a human being?," consider the following. It is very plausible to say, in answer to the question "What makes something a substance?" that the reason why, for instance, Socrates or his type of body is a substance, namely an independent reality, and not simply a collection of more basic material parts which are themselves independent realities, is because Socrates' body, his collected material

[16] *De An.* 2.1 gives us the key to understanding what Aristotle means by the "matter" (*hulē*, 1041b5) which is a *this* in Z.17 (1041b8). He uses the term "matter" in *de An.* 2.1 also, at 412a19, for the very thing that is made a substance and a *this* by the presence of form, and as such the "matter" is understood to be, paradigmatically, the familiar sort of natural living body with the capacity for standard life activities (412a19ff. with a8–9). Some have found this problematic since the matter, so construed, is also described by Aristotle in *de An.* 2.1 as a compound of matter and form (412a16). How, it is asked, can it be a compound of matter *and* form if the matter or material component of this compound itself already includes form? The answer to this question is that Aristotle uses the term *matter* (*hulē*) in this chapter in two ways just as he explicitly does elsewhere, e.g. at *GC* 1.4.320a2–5. In one use, matter is what is *potentially* informed (412a9), in another use it is what is actually informed (412a19). The potentially informed matter is indeed only a component of the compound or that from which it comes. The actually informed matter, however, *is* the compound, and is a *this* by virtue of its possession of form. Of course, as Aristotle makes clear in *de An.* 2.1, in the order of scientific inquiry one needs a way of first identifying the parcels of "matter" that are in fact compound substances in a way that permits one genuinely to discover what it is that explains why they are substances or *thises*, just as in the case of inquiry into thunder one needs a way of identifying the appropriate noises that belong to clouds in a way that permits one to discover why they so occur. To identify them initially as noises of the quenching of fire in clouds would pre-empt this effort. Rather, we initially identify them, paradigmatically, as the noises of a familiar sizzling or crackling sort in the clouds (see *APo.* 2.11.94b33). Similarly, substances are initially identified as the things that most of all present themselves as subjects having independent existence on their own such as, paradigmatically, the natural bodies of the sort that have certain basic life functions. Later, we discover what feature of them it is by virtue of which they have this independent status. We do not initially identify them as parcels of informed matter.

[17] For further defense of this reading of Z.17, in the context of Z as a whole, see Bolton 1995. On its further implications see Bolton 2002.

parts, have form, where form is understood in a functional or teleological way, as Z.17 suggests it should be (1041a28–30).[18] That is, Socrates' material parts function together in a systematically organized way to achieve certain ordered *goals* and a certain ultimate *goal*. At least, this is a very plausible answer to the question "What makes Socrates a substance?" by Aristotle's own lights, as we can see, for example, from Z.16 (1040b5–16). But is it plausible to say that this same type of account gives us a good or proper explanation of why Socrates is a member of the human species or of why any living thing is a member of its species?

There is one basis offered in the literature for thinking that for Aristotle it is. According to some, a main driving force behind Aristotle's doctrine of substance in *Metaphysics* Z is a new theory of predication which introduces significant revisions in the theory of predication of the *Categories*. According to the *Categories*, a biological genus or species, such as the human species, is a secondary substance that is *predicated* of its members, which are primary substances (see *Categories* 2, 5). In the terminology of the *Categories*, a genus or species is *said of* its members (which implies that the *definition* of the genus or species is also predicable of the members). In the *Categories*, so this story goes, this predication is regarded as ontologically basic in the sense that it is not reducible to any more basic ontological predication or predications. But in the *Metaphysics*, it is argued, Aristotle introduces his hylomorphic analysis of natural objects, including biological species, which he takes over from the *Physics* and other works, and he applies this to produce a new analysis of the predication of a biological species or genus of its members. According to this new analysis, it is, for instance, the predication of Socrates' *form* of his matter that is responsible for the fact that he is a member of the human species. This is the new explanation in the *Metaphysics* of why Socrates belongs to the human species, and in this explanation the primary *cause* or explainer of membership in the human species is *form*. In general in the *Metaphysics* and in Z.17 in particular, it is claimed, Aristotle wants to *explain* the membership of a particular primary substance (as understood in the *Categories*) in its biological species or genus in terms of a predication of form of matter.[19] If so, then, in this case, once again, it seems clear that the autonomy of biology and its separateness from metaphysics is threatened.

Now, as we have already seen, it is doubtful that any basis for this new view of predication is to be found in Z.17, since Aristotle is not there attempting to explain membership in a biological species or genus but

[18] For a recent defense of this see Charles 2000.

[19] For this proposal see, e.g., Lewis 1991, ch. 7.

rather membership in the class of substances, or of perceptible substances.[20] This, however, would not show that Aristotle does not articulate such a doctrine elsewhere. But there are also strong reasons to doubt that Aristotle wants to defend this doctrine elsewhere. On this doctrine, the predication of a biological species or genus of a primary substance, as understood in the *Categories*, is reduced, in the *Metaphysics*, to a predication of form of matter. But, of course, according to the *Metaphysics* there are various substances and living things that easily satisfy the main requirements of the *Categories* for being primary substances which are immaterial, such as the prime mover and the other divine substances of Λ.8 (1073b10ff.). In *Metaph.* Z.1 where he introduces the framework of the *Categories* Aristotle gives as examples of substances that are primary "a man or a god" (1028a10–20, 25–31). The predication of their genus or species of gods, however, as, for instance, when we say that the prime mover is a god or a specific type of god, obviously cannot be reduced to any predication of form of matter, since the prime mover and the other divine substances are immaterial. So Aristotle cannot mean to reduce all cases where a species or genus of living thing is predicated of a primary substance, as understood in the *Categories*, to predications of form of matter in the *Metaphysics*.[21]

On the other hand, it is true, of course, that Aristotle indicates, plainly enough, in the *Metaphysics*, that the principles and causes of *perceptible* substances (as opposed to immaterial substances), are just two, soul and body, or form and matter. As he says in Λ.5:

> These causes [of perceptible substances] are, doubtless, soul and body, or intellect and desire, and body. (1071a2–3; cf. H.1. 1042a24ff.)

This means, as we have seen in Z.17, that appeal to just these two principles, and primarily to form, is sufficient to account for the being of perceptible substances as such, namely as perceptible substances. But Aristotle also denies in *Metaph.* Λ.5 that appeal to just these two principles, form and

[20] Aristotle does mention the question "What is a human being?" at Z.17.1041b6ff., in order to compare how one should look for an answer to this question and to the question "What is substance?" But he does not *answer* the former question in Z.17. The only question he answers, at 1041b7–9, is "What is substance?" As we have seen, he does not say that *form* is what a human being is, but that it is what substance is.

[21] For similar reasons it cannot be the case in the *Metaphysics* that, at the basic ontological level, Aristotle wants to restrict the predication of substantial form or actuality, either specific form or more generic form, to its predication of matter, since the specific and generic form or actuality of any divine substance is irreducibly predicable of it and not of any matter. So no problem in the *Metaphysics* can depend for its solution on such a general restriction. Some have supposed, for example, that such a restriction is required for a proper understanding of Z.13 and that Aristotle relies on such a restriction to defend his view that no universal is a substance. This is problematic.

matter, is sufficient to account for why something is a member of a biological species such as the human species. He denies this when he argues that while form and matter are indeed the causes of perceptible substances they are not the causes of *all* types of things because they are *not* sufficient as cause (*aition*) for a human being. As he says:

> The cause of a human being is the elements, fire and earth as matter, and also the special form, and in addition something else outside, namely the father, and besides these things the sun and its ecliptic circle. (1071a13–16; cf. *Ph.* 2.2.194b13)

This point is essentially duplicated in *Metaph.* H.4 where the father's *sperma* is cited as the efficient cause of humans (1044a32–b8). In these passages Aristotle means quite definitely to argue *against* the claim that human form, or form and matter, alone are sufficient as causes to explain membership in the human species and he means to assert that the human efficient cause, namely the father, or the father's *sperma*, is required in addition, for a causally sufficient explanation. This implies, among other things, that if any entity should turn up that had human form, or even human form and matter, but not the human efficient cause, it would not be a member of the human species. This agrees with Aristotle's claim in *de An.* 1.1 that *each* of its *four causes* is a part of the definition and essence of – is constitutive of the being of – a natural object. Form alone, he argues there, is not sufficient, nor are form and matter (403a25ff.). So Aristotle does not attempt to reduce species membership to the predication of form of matter, in Z.17 or elsewhere.

It is clear, then, that Aristotle's discussion in Z.17 does not violate his requirement of autonomy and separateness for metaphysics and biology. The fact explained in Z.17, namely that Socrates or his type of matter is a *this*, is not a fact of biology but of metaphysics, and that Socrates' form is primary in this explanation does not show that it will be primary in the same way in the explanation of the genuine fact of biology that Socrates has two feet, and so forth. Here, his efficient cause, his father's *sperma* and its productive action is an essential cause.

A POSSIBLE RESPONSE

Now, I can conceive that someone might respond to this by claiming that while reference to the efficient cause might be required for a full *biological* explanation of membership in the human species it is not required for a *metaphysical* explanation of membership in the human species. For this form, or form and matter, are sufficient. This response, it should be noted first of all, would concede the *autonomy* of biology with reference to metaphysics which I have been urging on Aristotle's behalf, so that would no longer be at issue. But the response is still unsatisfactory. It is not only that in *Metaph.* Λ.5 and H.4, as in *de An.* 1.1, Aristotle specifically argues that form, or form and matter, are not sufficient to explain membership in the human species, without any qualification of the sort one would expect if he did take the view suggested. It is also that, at a quite deep level, this response begs the question. The very point at issue now is whether asking "What is it to belong to the human species?" is in *any* sense to ask a metaphysical question at all, or rather only a biological question. One needs some evidence that Aristotle is prepared to consider the idea that there are different answers to the question "What makes something a human being?," one biological and the other metaphysical.

There is one passage in particular to which one might appeal to provide such evidence, namely the well-known passage already referred to in *de An.* 1.1 where Aristotle draws distinctions between various different types of definitions of the same entity. He points out that a natural object such as anger "would be defined differently by the *dialectician* and by the *natural scientist*" (403a29). The context makes it clear that Aristotle means by this that one answer to the question "What is anger?" could be quite adequate from the dialectical point of view (i.e. *kata doxan*) and quite inadequate for natural science (i.e. *kat' alētheian*) where different requirements are in force (cf. *APo.* 1.19.81b18ff.). There is no reason to doubt that Aristotle would apply this same distinction to the definitional question "What is a human being?," since answers to this same question can be treated by dialectic and by natural science. The tests for the adequacy of a dialectical answer to this question are, of course, mainly laid out in the books on definition in the *Topics* (6–7), while, as we have noted, the very different tests for the adequacy of a scientific answer to this question are laid out in book 2 of the *Posterior Analytics*, and in various methodological passages in the scientific works.

By analogy with this, some have suggested that just as there are proper dialectician's *and* physicist's answers to the questions "What is anger?" or "What is a human being?," so there are proper metaphysician's answers to these questions.[22] If the natural scientist's answer to this question cannot be in terms of form alone, or form and matter alone, perhaps the metaphysician's answer can. But the text of *de An.* 1.1 does not say this and

[22] See, e.g., Frede 1990.

what it does say relevant to this topic tells against this. Aristotle goes on to assert in this passage that just as the physicist deals with entities that are inseparable from matter both in their scientific definition and in fact, and just as the mathematician deals with entities that are inseparable in fact but separable in definition (or by abstraction) from matter, so the metaphysician deals with entities insofar as these are separated in fact (and thus also in definition), from matter (403b14ff.). This duplicates, of course, distinctions which Aristotle draws in *Metaph.* E.1, where he again contrasts the different types of theoretical science and says that while the mathematician deals with things which are changeless (so far as their definitions are concerned) but do not exist separate from matter, the primary branch of philosophy, metaphysics, deals with things which both are changeless and exist separate from matter (1026a6ff.). But just as these distinctions, as found in these passages, rule out any *mathematician's* definition of anger or of the human species, since these objects are not, according to their definitions but not in fact, changeless and separate from matter, so these distinctions do not leave room for any *metaphysician's* definition of anger or of the human species, since these objects do not exist separate from matter. So while *de An.* 1.1 does leave room for both a dialectician's and a theoretical scientist's definition of anger or of the human species, neither it nor *Metaph.* E.1 leaves any room for a mathematician's or a metaphysician's definition. The only possible *scientific*, as opposed to dialectical, definition or statement of essence of the human species is assigned in *de An.* 1.1 to the *natural* scientist. So the only possible scientific answer to the question "What makes something a member of the human species?" must come from the biologist, not the metaphysician, according to these texts.

As we have already noted, *Metaph.* E.1 does, of course, add, as *de An.* 1.1 does not, that metaphysics, or what is called in E.1 theology, does include the study of everything that is *qua* being (1026a23–32). From this it follows that metaphysics does indeed include the study of anger, and also of human beings, *qua* being, that is, *qua* their appropriate mode of being. That is, it includes the study of anger *qua* quality and of human beings *qua* substance. So metaphysics *will* answer the question of what it is that makes the wrath of Achilles or any other instance of anger a quality, not a substance or a relative. And metaphysics will answer the question of what makes Socrates, or any other living thing, a substance, not a place or a time. It will answer the latter question by appeal to the fact that Socrates has a certain sort of form or soul; and it will answer the former question similarly, by appeal to the fact that the wrath of Achilles is an affection subject to more and less intensity or something of that sort.[23] But these answers will not be metaphysician's answers to the questions "Why is Socrates a human being?" or "Why is the wrath of Achilles anger?," since they are not answers of any sort to these questions. So Aristotle's answers to the questions "Why is the wrath of Achilles a quality?" or "Why is Socrates a substance?," as a metaphysician, provide no basis for a metaphysical analysis of the proposition that Socrates is a human being or of the proposition that the wrath of Achilles is an instance of anger.

WHAT KIND OF FORM IS A SUBSTANTIAL FORM?

This line of thought has important implications for one other issue mentioned earlier, one much discussed in recent years, where it has again been supposed that biology is an essential source for Aristotle's metaphysical doctrine. This issue concerns whether substantial forms for Aristotle – that is the forms that are responsible for the substantiality of substances – are species forms such as human form or rather are sub-species or particular forms such as those which, some claim, are invoked in *GA* 4.3 to account for inherited family resemblances. Since, on the line taken by Aristotle in *Metaph.* Z.17, substantial forms are only invoked in metaphysics to explain the common defining attributes which substances have as substances – such as being a *this* – there is no need for and no role in metaphysics itself for any sub-species forms, or other entities, which might be needed to account for attributes which belong to a substance, say Socrates, not as a substance but as a member of a certain family. As we have seen, metaphysics, like any science, only invokes in its principles entities that are required to explain its special data, and shared family resemblances are not among the data of metaphysics since they have nothing to do with what is *qua* being. Equally, however, for the same reasons, it is not clear that metaphysics needs to invoke, as substantial forms, species forms either. For just as the attributes which substances have as substances are not attributes which they have as members of families, so they are not attributes which they have as members of any particular species of living thing. As *de An.* 2.1 makes clear, the possession of soul or life form in general is sufficient for Aristotle to explain why something is a *this*. So it is at that level that explanation of *thisness* takes place, namely at the level of the *generic* form of living thing and not

[23] In *Cat.* 8.10b26ff. Aristotle discusses this and other suggestions as to what it is that is distinctive of qualities as such.

at any more specific level. Moreover, since the generic life form in question is one which, for Aristotle, belongs to immaterial, divine substances as well as to perceptible substances, the form in question, substantial form, is one whose scope of application goes well beyond the realm of biology or physics.

The situation here may be usefully compared with another to which Aristotle often draws attention, for instance in *APo.* 1.4–5. As he makes clear there, one does not need to invoke, and in fact one must not invoke, the form or essence of the isosceles triangle to explain why isosceles triangles have interior angles equal to two right angles, because this property is a property that isosceles triangles have not as isosceles triangles but rather simply as triangles. So the only form and essence which needs to be theoretically invoked to explain this property of isosceles triangles is the generic form and essence of the triangle. Of course, for Aristotle, there is nothing which has the generic form of a triangle that does not also have the form of some specific type of triangle such as the isosceles triangle. Also, for Aristotle, there is nothing that has the form of the isosceles triangle that does not also have features that make it a particular isosceles triangle. Contrary to Plato, for Aristotle there is no entity which is just a triangle or just an isosceles triangle without being a concrete particular triangle (*Metaph.* Λ.5.1071a17ff.). But this does not mean that Aristotle can dispense with the generic form of the triangle in geometry and do all of his explaining with more specific forms or features. He clearly rejects this in *APo.* 1.4–5. Equally, in metaphysics, anything that has generic substantial form will, as a living thing, also have some species form, and perhaps also one or more types of sub-species form as well. But none of these sub-generic forms, such as there may be, will be forms of the right type for Aristotle to explain the defining attributes of substances as substances, such as being a *this*, since none of these defining attributes of a substance is species specific, or sub-species specific. The scope of the debate, then, over whether substantial forms for Aristotle are species forms or sub-species forms needs to be expanded to include a third possibility – namely that substantial form is supra-species form – since this possibility fits best with certain main requirements of Aristotle's views on scientific explanation.

Of course, any attempt at a defense of this third option would have to find a way to deal with certain problems. In particular, as specialists on *Metaphysics* Z will quickly remind us, it will have to deal with Z.13 and related texts where Aristotle argues that, in some sense, the universal and the genus are not substance. There are now in the literature various well-known strategies for attempting to show that the argument of Z.13 does not rule out species forms as substantial forms even though species forms are, in an ordinary sense at least, universals which belong in common to many things. One or more of these strategies could be used to show that Z.13 does not rule out supra-species form as substantial form either. The simplest of these strategies, perhaps, is based on the view that Z.13 does not rule out anything because it is merely aporetic, and inconclusive (see 1039a14ff.).[24] This is not a claim, however, that I wish to try to defend. All that I wish to argue here is that it is not possible to determine which of these three ways of understanding substantial form is correct by anything proper to Aristotle's biology since the entities invoked for explanation there are not invoked for the explanation of any strictly metaphysical data. By contrast, Z.13 introduces strictly metaphysical considerations which are entirely proper to the resolution of this issue.

It might still be argued, as some have, that there *is* an explanatory role for sub-species forms in metaphysics itself on the grounds that it is the proper job of metaphysics to explain not only why something is a *this*, namely a numerically single independent reality, but also to explain why any given *this* is the particular *this* that it is. For such a purpose, some would argue, sub-species forms are required, so they are needed in metaphysics. In response one should say, first, that such sub-species forms, should they be needed, would still not be substantial forms. That is, they would not be the forms needed to explain why something is a substance or a *this* but only to explain why some substance is the particular substance or *this* that it is. These are different explanatory projects, and entities of the appropriate type and generality would be needed for each project, just as the project of explaining why some concrete particular isosceles triangle has angles equal to two right angles is a different project than that of explaining why that triangle is the concrete particular isosceles triangle that it is. The generic form of the triangle is what is needed according to Aristotle for the first project but it is obviously not sufficient for the second. In addition, according to a standard reading of *Metaph.* Z.8 at least, what explains why two substances such as Socrates and Callias are the different particular substances that they are is not that they have different sub-species forms but that they have different matter, while in form they are the same (1034a5–8). Still, this standard account of Z.8 and of its significance is itself disputed, and this is another issue that cannot be settled here. But, once again, this permits us to see that however it is

[24] See, e.g., Burnyeat 2001, 44–52. Other strategies are discussed in, e.g., Lewis 1991, ch. 11. See also Bolton 1995.

to be settled the relevant evidence for settling this strictly metaphysical issue must itself be strictly metaphysical and not essentially dependent on any material proper to biology or on the explanation of any strict facts of biology. So this illustrates again the importance of seeing the ramifications of Aristotle's insistence on the autonomy of metaphysics and of biology for our understanding of his doctrines and projects in each area.

CHAPTER 3

The unity and purpose of On the Parts of Animals *1*

James G. Lennox

INTRODUCTION: THE UNIT OF UNITY

In the first chapter of *Generation of Animals* 5 Aristotle turns to a discussion of the affections in virtue of which parts differ from one another, noting that while some of these are found universally throughout kinds of animals, some (e.g. color of eyes, hair, feathers) are distributed somewhat randomly, and develop at different times in different individuals. "We should not suppose," he cautions, "that the mode of explanation in these and all such cases will be the same" (778a29–31). In particular, if a feature is neither common to animals nor a distinguishing feature of some kind or other, we should not suppose it either is or came to be for the sake of anything. Rather, in these cases we should suppose the affections in questions came to be *of necessity* and should explain them by reference to *matter* and *the source* of their having changed. He then seeks for a deeper justification.

For as was stated according to principles in the primary accounts (*kat'archas en tois prōtois logois*), it is not because each thing's coming to be is of a certain sort that it *is* of a certain sort; rather it came to be such a thing on account of its *being* of this sort; for generation follows on and is for the sake of being; being is not for the sake of generation. (778b2–7)

In the context of justifying a discussion of attributes of animals that are said not to have come to be for the sake of an end, it is not entirely clear how invoking this principle helps to clarify what he has just said, but

This chapter was first a paper delivered in October of 2004, at a conference organized by Robert Bolton and me to celebrate Allan Gotthelf's contributions to Aristotle scholarship, held at the University of Pittsburgh shortly after Allan was appointed Visiting Professor in the Department of History and Philosophy of Science. I first met Allan in my senior undergraduate year, and for the next decade he served as a valuable and generous mentor, a role that gradually transformed as I became a "co-conspirator" in encouraging scholars of ancient philosophy to take Aristotle's scientific study of animals more seriously as a valuable resource for deepening our understanding of every aspect of his philosophy. It is a great pleasure to be able to contribute to, and serve as editor for, this volume in Allan's honor.

in this essay that is not my concern. Rather, I am interested in what this passage may tell us about the unity and structure of Aristotle's investigation of animals, and its place in his wider investigation of nature.

It is reasonable to assume, as commentators typically do, that the reference to what was stated according to principles in the "primary accounts" is to the following lines of *PA* 1.1:

For even with house building, it is rather that these things happen because the form of the house is such as it is, than that the house is such as it is because it comes to be in this way. For generation is for the sake of being; being is not for the sake of generation. (640a15–19)

Supposing this is the statement being referred to – in what sense are the *logoi* of which this is a part *primary*, and *to what* are they primary? One answer was suggested in I. Düring's commentary to *De partibus animalium* (Düring 1943). Düring developed an elaborate theory of a course of lectures on animals that grew and developed over Aristotle's career, to which *On the Parts of Animals* 1 was a late addition, intended as an introduction to the "third course of [zoological] lectures." He assumed the extension of *logoi* in our *GA* 5.1 passage was to the whole of a "lecture" corresponding to our *PA* 1; he sometimes writes as if one should think of the plural as referring to five distinct but related parts of this lecture.[1] As assumptions, these seem rash. The phrase *en tois prōtois logois* and its companion *kata tous prōtous logous* are stock phrases in Aristotle. To illustrate the dangers of loading the phrase with too much baggage, two uses in the *Politics* will serve. At 1278b18 and 1325a30 the references can be taken, with some confidence, to be to book 1 in our modern editions, and indeed the first reference is to the very opening arguments of book 1 – "first" in these cases means something like the discussion that comes at the beginning of our *Politics*. But the same phrase appears at 1288a37, and there it apparently refers to a passage earlier in the same chapter; while at 1295a4 it refers to an earlier book, but decidedly not our *Politics* 1.

Nevertheless, the importance of this phrase as it is used in *GA* 5.1 is that Aristotle is apparently referring to some part, or perhaps all, of our *PA* 1 – that is, to a text *outside* his study of animal generation – that discusses the philosophical question of how best to conceive of the relationship between the process of coming to be and the being that results from that process. It therefore provides independent support for the idea that the

[1] Düring 1943, 36: "The lecture consists of the following parts." Balme refers to it as "a collection of five separate papers" but also notes that the "topics are discussed in a reasonable order . . . and are all necessary," Balme 1992, 69.

first book of *On the Parts of Animals*, or some part of it,[2] was intended as a philosophical prolegomenon to the entire investigation of animals, and not merely to the three books on animal parts to which, since Roman times, it has been attached. This in turn invests *protoi* with a degree of normative force – not just to something that came earlier in a sequence, but to something that sets out *philosophical primaries*. For, as I will argue, a central theme running through *PA* 1, and which binds it together, is that there are important methodological implications that derive from the fact that animals are natural substances that are produced by a very special kind of unqualified coming to be, directed by, in Gotthelf's phrase, an irreducible potential for form. Thus, questions about the relationship between the study of actual, mature animals, on the one hand, and the study of the natural process of generation and development that produces them, on the other, are prominent in *PA* 1.1; and the answers to those questions have implications for how Aristotle imagines the relationship between the investigations represented by *Generation of Animals* and *On the Parts of Animals* 2–4.

This reference in *GA* 5.1 to a principle established in *PA* 1.1 that ought to govern how we approach the study of animal generation also encourages us to think of the study of animals as having an internal coherence that to some extent sets it apart from other contributions to the study of nature.

Meteorology 1.1 supports this suggestion. After a review of other parts of the science of nature, parts that correspond in a general way to our *Physics*, *On the Heavens*, and *Generation and Corruption*, Aristotle argues that meteorology is "a part of the same *methodos*" (338a26).[3] After reviewing the topics to be taken up by meteorology, Aristotle goes on:

> Having dealt with these subjects, we will study whether we are somehow able, according to the recommended manner,[4] to give an account of animals and plants, both in general and separately. For having given an account of these things we would pretty much have reached the goal of our original plan in its entirety. (339a5–9)

[2] There are a number of good reasons to think that the chapters 2 and 3, on division, were originally independent. However, the last paragraph of chapter 4 is a summary of everything that precedes it, including the discussion of division; its author thus recognizes a common project covering the first four chapters. There are also good reasons to think that chapter 5 was also originally an independent *logos*, or perhaps two independent *logoi*. But I will have much more to say about it later.

[3] For *methodos* in this sense, see Pl. *Sph.* 218d, 235c; *PA* 1.1.639a1; 1.4.644b15; *EN* 1.1.1094a1. It can also refer to a shared *doctrine*, as at Pl. *Tht.* 183c2; and to the particular *method* used in an inquiry, as at *Pol.* 1.1.1252a18, where it introduces a method to be used both in politics and "in other *epistēmai*" (cf. *Pol.* 1.1.1289a18, cf. *EN* 5.1.1129a6).

[4] The Greek is *kata ton huphēgēmenon tropon*; I have tried to translate *huphēgēmenon* to capture the implicitly normative force it seems to carry in these contexts.

Notice that there is both a meta-level inquiry that is referred to here, as well as first-order inquiries into animals and plants. The first question is whether such inquiries can somehow be pursued in the manner proposed for the other subjects. Supposing an answer to that question, we can then go on to study, both generally and distinctly, animals and plants. Finally, the concluding sentence indicates that this study has all along been seen as a fitting completion to the planned investigation of nature.

I will now turn to outlining a case that a central motivation for Aristotle in writing book one of *On the Parts of Animals* was to address one facet of this question.[5] Two features of the way *PA* 1 is framed must be kept in mind as we proceed. First, the investigation is framed not as exploring methods for an inquiry into *animals*, but for an inquiry into *nature* (cf. 639a12–14). Nevertheless, it is clear that its primary aim is to specify and supplement general principles for the *specific* purpose of studying a special sub-category of natural entities, those that undergo an especially complex form of unqualified coming-to-be and passing away.[6] The nature of this substantial coming-to-be demands that, among the causes studied by the natural scientist, goal-causation should be primary, and the form of necessity invoked must reflect that primacy.[7] Moreover, while it is generally true that the natural scientist must investigate both the material and formal nature, in this sub-category of natural being the formal nature is *soul*. And finally, while in any science it will be requisite to make use of demonstration and division, Aristotle here insists on a *special kind* of demonstration and division.

Seeing the second order question in *Mete.* 1.1 as establishing the agenda for *PA* 1 provides a useful way of envisaging the unity of purpose of its various *components*. In this instance one might ponder how different coherent arguments are related to various other works that treat of similar topics. Thus I have argued on another occasion that virtually every topic discussed in these five chapters bridges a gap between the basic framework of scientific inquiry and knowledge found in the *Posterior Analytics* and

[5] This discussion thus provides detailed support for the suggestive comments about the place of the study of animals within the wider study of nature in Falcon 2005, 4–7 and Lennox 2005b. Cf. Burnyeat 2004 in de Haas and Mansfeld 2004, 13 n. 16, pointing to the possibility of the option for which I will here argue.

[6] It is not difficult to be precise about the nature of the complexity. For example, biological generation begins with a small amount of a single uniform part (a form of blood, menses, in blooded animals), and over a period of days, weeks or months, depending on the species, what comes to be is a coordinated system of functioning non-uniform parts.

[7] That is, the primary contrast, which comes out explicitly in chapter 5, is between a study of *eternal* natural things governed by *absolute* necessity, and *generated* natural things governed by necessity *conditional on an end*. Compare *PA* 1.1.639b6–640a9 with *PA* 1.5.644b22–645a6.

the needs of an investigation of natural substances subject to unqualified coming-to-be and passing away.[8]

Here, however, I shall make the case for a different sort of unity, which I will refer to as *narrative* unity – that is, a unity that builds gradually toward the standards of investigation and proof that Aristotle is seeking, and reaches what we might call a climax in chapter 5. If this is a metaphor borrowed from the genre of tragedy or the novel, it can nevertheless be instantiated in quite rigorous ways. Euclid's *Elements*, for example, displays at least some of the features of the narrative unity that I will attribute to *PA* 1. In particular, both of them are constructed in such a way that the arguments later in the treatise depend, in complex ways, on conclusions established by earlier arguments.

LOOKING BACKWARD

In "Division and Explanation in Aristotle's *Parts of Animals*" Allan Gotthelf noted that the part of chapter 5 of *PA* 1 that immediately follows the well-known encomium to the study of biology twice refers back to ideas and arguments found in chapters 1 and 4;[9] and that chapter 4 tentatively responds to an unanswered *aporia* that is developed at length in chapter 1. Moreover, chapter 4 defends a method for identifying kinds at a certain level of generality, and these kinds are apparently presupposed by the reformed theory of division defended in chapters 2 and 3. All of this suggests a rather tightly integrated discussion.

On the other hand, *PA* 1 is regularly characterized as *lacking* any overall sense of unity and purpose. These characterizations are sometimes supported by statements about radical differences in style from one section to the next and by conjectures about parts being written for different purposes or at different periods of Aristotle's philosophical career. There is no doubt that the stylistic differences between the exhortation to the study of animals in 644a22–645a36 and (to pick the most obvious contrast) the attack on dividing dichotomously in 642b5–643b8 are stark. And there are signs, such as the conclusion of chapter 4 which seems not to look forward to the material in chapter 5, that this book may have been put together from material written at different times and for different occasions. I will argue that the somewhat disjointed structure and disparate styles of *PA* 1

[8] Lennox 1996, 239–47; reprinted in Lennox 2001a, ch. 4.
[9] Gotthelf 1997a, 215–30.

do not conflict with the evidence for tight integration I outlined in the previous paragraph.

To understand the nature of that integration, I will need to say a bit more about the notion of "narrative unity" and argue that this is the way in which that integration is achieved. By "narrative unity" I have in mind the sort of unity that is referred to as "plot" when one is discussing fiction writing. In the case of *PA* 1, we will see that there is an order in which its questions are introduced that is mysterious until one progresses further into the "story"; some questions are raised and left unanswered until later, sometimes much later, after materials needed for a proper answer have been developed. The idea that the student of this sub-category of natural substance will, in studying form, be studying soul is introduced late in the argument, after first rejecting a straightforwardly materialist approach, and then rejecting "structuralist" theories of form (640b4–641a33). An account of multi-differential division emerges gradually as a response to problems in a theory that uses only dichotomous division (642b5–644a11); and so on. One more feature is characteristic of narrative unity: once a certain position has been established, argument will proceed on the assumption that the reader has taken that position on board. Later arguments will look incomplete on their own, because they depend on the reader's assuming a number of premises that go unmentioned. Like reading a plot, one cannot jump around randomly in the text and hope to fully understand what is going on.

Having said that much, I will now present in outline a case for just this sort of internal, narrative unity in *PA* 1. At first, given what I have said, it may seem odd that I begin by focusing on the fifth chapter. However, I do so in order to show how much in the previous chapters is presupposed by it. Moreover, by tracing back from chapter 5 I can also make the case that, while consisting of two very different sections, this last chapter constitutes a unitary capstone for the work as a whole.

To a certain extent this is a challenge to the position I argued for in 2001.[10] At that time it seemed to me that, while the first half of chapter 5 was a rhetorically powerful call to philosophers to take up the study of animals, in was an intrusion. Today I argue that, while the earlier chapters of *PA* 1 constitute a narrative with many, complexly intertwined, arguments developing simultaneously, chapter 5 provides, as it were, the vision of the investigation of animals that emerges from that narrative and locates

[10] Lennox 2001b, 174.

that investigation relative to other, more familiar, investigations of natural things.

We begin with the second half of the chapter, running from 645b1 to the end of the book 1. Aristotle is integrating two strands of the argument of the first four chapters that, up to this point, have remained entirely detached. Much of chapter 1 is engaged in developing a model of teleological explanation and conditional necessity, while chapters 2–4 are engaged in developing a multi-differentia method of division, which leads directly into a discussion of what kinds are appropriate for such division and how they are to be identified. How the results of these two discussions are related is mysterious.

Allan Gotthelf began "Division and Explanation" by making the case that we should render *dielein* at 645b1 in its technical sense, "to divide," as do Balme and Pellegrin. He translates:

> It is necessary first to divide the attributes pertaining to each kind which are present in all the animals [of that kind] in virtue of themselves, and after that try to divide their causes. (645b1–3)[11]

His case rests in part on the fact that the next few lines refer us back, almost certainly to discussions in chapters 1 and 4, where the context is "divisional," and in part on the patterns of actual explanations in *PA* 2–4. The next few lines are:

> Now *it has been said before* that many common features belong to many of the animals, some without qualification (such as feet, wings, and scales, and affections too in the same way), and others analogously. By "analogously" I mean that while some have a lung, others have, not a lung, but instead something different which is to them what a lung is to those that have one; and some have blood, while others have its analogue, with the same potency that blood has for the blooded. To speak separately about each of these animals as particulars, *as we also said before*, will result in saying the same thing many times, whenever we speak about all their attributes; the same attribute belongs to many animals. (645b3–14)

The first backward reference is probably to the discussion in chapter 4, where Aristotle takes up the question of how to identify kinds at appropriate levels of universality for investigation. He begins simply by stipulating that it is correct that those animals with parts that differ only in degree or by more and less should be designated as kinds, while those with features that are analogous have been "kept apart" (644a15–22). He then reintroduces a problem first identified in chapter 1, and now presented as an *aporia*: in

[11] Gotthelf 1997a, 222.

so far as what we want to understand is the being of the indivisible forms in nature, it seems best not to study kinds in general, but particular forms of the kind; but "in so far as this will result in speaking many times about the same affection because it belongs in common to many things, in this respect speaking separately about each one is somewhat silly and tedious" (644a32–34).[12]

The resolution of the *aporia* follows immediately. We should speak in common about kinds when they "have a single common nature and forms within them that are not too distant" (644b3–4), even in cases where the *genos* has yet to be named. Now he does not here say anything more about this "single common nature"; but just prior to this discussion Aristotle has said that people have acted appropriately in grouping together, within a kind, forms whose parts differ only by "more and less" or "excess and defect" – that is, parts that differ only in degree along various continuous dimensions. It is plausible, then, that by "forms not too distant" he means animals whose parts can be ranged along such continua. In that case, then, we have evidence of "a single common nature" if all the divisions of a kind stay within the bounds of more-and-less variation.[13]

When a group of different animals meets these conditions, this constitutes reasonable grounds for belief in the existence of a kind with a common nature. This belief can be formed prior to knowing – and will serve the aim of knowing – what the common nature is.

And now it is critical, for seeing the integrative role of chapter 5, to recall that the other central thread of *PA* 1 is a sustained argument for three *very different*, interrelated claims: (i) that it is the organized biological capacities of an animal, its "soul," that constitute an animal's nature and being (see especially 641a14–32); (ii) that the soul is that for the sake of which (the final cause of) the differentiated tissues and organs of an animal come-to-be with the structure, character, and interrelations they have (639b12–21; 640a10–32); and (iii) that there is a special sort of necessity, "conditional necessity," that grounds teleological demonstrations relating differences in parts to differences in living function (639b22–640a9; 642a1–13). To have identified a kind by the standards we have just discussed is a far cry from knowing *why* there is a kind so constituted, or *what* the nature of that kind

[12] See 639a16–27. Compare *APo.* 1.5.74a25–32; and 1.24.85a20–b15 on the same issue in the context of the theory of demonstration. I make the case that the concern about repetition is non-trivial in Lennox 1987a, 114–18 (reprinted in Lennox 2001a, ch. 1). How deeply this concern runs through Aristotle's metaphysics and epistemology can be seen by the fact that it is the final *aporia* of *Metaph.* B (1003a7–17) and by the fact that much of *Metaph.* Z is concerned with a tension between the greater reality of the particular and the greater epistemic value of the universal.

[13] I have pressed this interpretation in more detail in Lennox 2005a, 87–100.

is. Moreover, much of the work of biological investigation has to do with what I have called the divisional articulation of these basic kinds into their differentiated sub-kinds and forms. But none of this can be accomplished without the preliminary task of *fixing* the kinds being completed. That is, the method for the investigation of causes outlined in chapter 1, together with the method of division defended in chapters 2 and 3, presupposes a way of establishing the kinds to be investigated; while the method of establishing those kinds is incomplete and tentative, without a way of getting at their natures and a theory about how those natures explain the varying differences that gave us a grip on the kinds in the first place.

But to show that these two aspects of animal investigation are related is, I will now show, the integrative task of chapter 5; for immediately after reminding us of what has been accomplished in chapter 4, Aristotle goes on:

> Now since every instrument is for the sake of something, and since each of the parts of the body is for the sake of something, and that for the sake of which is some activity, it is apparent that the entire body too is constituted for the sake of a certain complete activity. (645b15–18)

The passage concludes by saying that the body is somehow for the sake of the soul, and each of the parts is for the sake of the function in relation to which it is naturally suited. Those continuously varying parts and bodies, then, that chapter 4 argued gave us our first fix on kinds with common natures, will be explained teleologically by reference to the activities they perform. But how will we be able to identify the appropriate activity? To deal with that question, the machinery of similarity and difference that was developed to talk about how appropriately to group and differentiate the parts present in kinds is applied to the *activities* for the sake of which these parts are present.

> Therefore we should speak first about the activities [of animals], those that are common, those that are spoken of according to kind and those that are spoken of according to form. By "common" here I mean activities that belong to all animals; by "according to kind" I mean those whose differences from one another are in degree, for example I speak of bird according to kind, but of mankind, and any other the general account of which has no differentiation, according to form.[14] Some activities they have common according to analogy, some they have according to kind, and some according to form. (645b21–8)

14 It is worth pointing out that the resolution of the *aporia* about the proper level of investigation in chapter 4 uses "bird" as the example of kinds with forms "not far apart" and uses mankind as the example of undifferentiated kinds to be studied "according to form" – that is, the same examples used here. Thus the discussion picks up on the earlier one in subtle as well as obvious ways.

Certain activities, say cooling, hold in general for all blooded animals, by analogy. Cooling of the heart is performed by the lungs taking in air in birds, four-legged land animals, and cetacea, while it is performed by gills taking in water in fish. The lungs of the animals that have lungs, on the other hand, differ in size, moistness, density, color, and so on, differences which Aristotle explains by reference to the relative moistness or dryness of the animal's body, warmness of its heart and blood, and its way of life. As 645b1–2 led us to expect, then, these divisions, which specify differences in the lungs of different animals are to be accompanied by divisions that specify the underlying causes of those differences.

The chapter then concludes by laying out a complex network of teleological and non-teleological relationships among parts and activities that can be diagrammed as below (→ = for the sake of):

1. $\text{Action}_1 \rightarrow \text{Action}_2$
2. $\text{Part}_1 \rightarrow \text{Part}_2$
3. $\text{Part}_1 \rightarrow \text{Action}_1$
4. Z is necessarily present because X, Y, . . . are.[15]

All of these causal relationships are repeatedly instantiated in the practice of *PA* 2–4. To stick with the same example:

1. Breathing, namely taking in and expelling air, is *for the sake of* cooling the heart region (669a5–6).
2. The neck is present *for the sake of* the larynx and esophagus (664a15–17).
3. The larynx is present for the sake of breathing (664a18–36).
4. The expansion and contraction of the lung is *necessary because* air is being taken in and expelled (669a16–17; cf. 642a35–b2).

In other words, the divisionally articulated activities and the divisionally articulated parts are now to be understood as *causally related* divisional articulations. In fact a good part of the work in Gotthelf's "Division and Explanation" is in identifying passages in the later books of *On the Parts of Animals* where there is a division of attributes found in a *genos* and explanations sought by means of a parallel differentiation of correlated functions – "first divide the attributes, then attempt to divide the causes."

Now despite the fact that the examples used in chapter 4 to exemplify between-kind and within-kind variation are variations in *parts*, we have been prepared since the beginning of chapter 1 for a parallel division of activities. For when the *aporia* at the heart of chapter 4 is first raised at the beginning of chapter 1, Aristotle distinguishes attributes that belong in an

[15] The chart is borrowed from the note to the passage in Lennox 2001b, 177. Explanation type 4 appears to stand in for the type of non-teleological necessitation referred to twice in chapter 1, at 640a33–b4 and 642a31.

undifferentiated way in different forms of a kind from attributes that are differentiated in form. The examples used to highlight this distinction are not parts but activities and passions – sleep, respiration, growth, and death exemplify attributes that are undifferentiated in form; while locomotion exemplifies an attribute that must be differentiated in form into flying, swimming, walking, and crawling.

There is, then, a compelling case for reading the last half of chapter 5 as a conscious attempt to integrate the two central themes of the first four chapters, the proper mode of causal explanation in the study of animals, and the proper way to identify and differentiate the kinds and differences to be investigated.

LOOKING FORWARD

Having made a case for seeing the last section of *PA* 1.5 as a deliberate integration of the methods for investigating teleologically organized unities of material and formal natures defended in chapter 1 with the methods of division applied to the general differences of properly identified and differentiated kinds defended in chapters 2–4, we can now look in more detail at the narrative development of chapter 1.

This chapter opens by pointing to a distinction of cognitive capacities that can be found in *every study and systematic inquiry* (*Peri pasan theōrian te kai methodon . . .*) – a capacity properly referred to as scientific knowledge (*epistēmē*) of the subject of investigation; and a capacity to discriminate between well-presented and poorly presented reports of the results of an investigation – a certain sort of learned skill (*hoin paideian tina*, 639a1–4). Among people who have this latter capacity, Aristotle introduces a further distinction, between those who are able to make such discriminations across a wide range of subjects, and those who can do so "about something of a delimited nature." He later claims that making such discriminations is *not* a matter of determining whether the claims being examined are true or false. Rather, using a certain set of standards, the *pepaidoumenos* regarding a determinate subject can discriminate well-formed and ill-formed claims independently of knowing whether the claim is true or false. For example, near the close of *GA* 5.8, Aristotle recounts Democritus' claims about the timing and mechanics of teeth emerging and falling out.

> So though nothing prevents the teeth coming-to-be and falling out in this way, nevertheless it is not on account of these [necessities], but on account of the goal; these necessities are causes as producing movement, as instruments and as matter . . . (789b6–8)

Nothing prevents Democritus' account of the mechanics being true; nevertheless, he has failed to see that these material changes are taking place for the sake of some end. Understanding one of the standards of judgment developed in *PA* 1.1 – that the "for-the-sake-of-which" cause has explanatory priority over the "whence-the-source-of-change" cause – gives the person with the proper training about the investigation of nature the ability to recognize the limitations of Democritus' theory, independently of whether what Democritus says is (as far as it goes) true.

Aristotle immediately tells us that in natural inquiry *too* we require certain standards (*horoi*) by which we can appraise the manner of its proofs; presumably it is by grasping such standards that one comes to possess the capacity of the *subject-specific*, discriminating reader/investigator. And indeed, the articulation of standards by which to evaluate natural inquiry begins with the very next sentence.

There is not space here to discuss each of these standards in detail. Rather, there are two tasks that can be accomplished, and which point both to the narrative unity of this chapter and its connection to those that follow. First, I will look carefully at the discussions of the first three standards. Each discussion begins by posing a question about an alternative; the first two are posed but not immediately answered; and the third, which *is* immediately answered, develops into a lengthy discussion that concludes with what appears to be an answer to the second question. Second, I will look at the first four chapters of *On the Parts of Animals* 1 as an extended discussion of the standards needed to make sound judgments about natural investigation.[16]

As already noted, the first is restated as an *aporia* in chapter 4, and it is revealing to compare its initial statement and its reformulation in chapter 1 with its aporetic reformulation in chapter 4. Here is the first formulation in chapter 1:

> should one take each substantial being (*ousia*) singly and define it independently, for example taking up one by one (*kath' hekaston*) the nature of mankind, lion, ox, and any other animal as well; or should one first establish, according to something common, the attributes common to all (*ta koinēi sumbebēkota pasi*)? (639a15–18)

I want to draw attention to two features of the wording here: the first option on offer is to take up and define distinct *ousiai*, which are then identified as the *natures* of specific kinds of animals; while the second option on offer is to take up *attributes* according to something common. After discussing certain pros and cons of the options, and introducing an

[16] Each is briefly discussed in Lennox 1996.

important distinction between general attributes that cannot be differentiated by division and others that can, he reformulates the options:

Accordingly, the following question about how one is to carry out an examination should not be overlooked – I mean the question of whether one should study things in common according to kind (*koinēi kata genos*) first, and then later the distinctive characteristics (*peri tōn idiōn*), or whether one should study them one by one straight away. (639b3–6)

The "something common" is now referred to as a kind (*genos*); the options are now temporally prioritized; the question is now whether one should study what belongs in common according to (or "in virtue of") kind *first*, and *then* study those that are distinctive to the specific forms of the kind; and finally, that alternative is now an option of studying "distinctive characteristics" (*idia*) – substantial beings and the natures of animal kinds are not mentioned.

As I have already mentioned, Aristotle passes on to the next question without attempting to resolve this one; but he returns to it in chapter 4. And he does so by reiterating a point from chapter 1 – that since it is the final forms (presumably in a division) that are substantial beings, if we do not speak first about what belongs in general to them, we will repeat ourselves constantly. He then presents the alternatives as an *aporia*.

On the one hand, in so far as what is indivisible in form (*to tōi eidei atomon*) is a substantial being (*ousia*), it would be best, if one could, to study separately the things that are particular and undivided in form (*peri tōn kath' hekaston kai atomōn tōi eidei*) – just as one studies mankind, so too bird, for this kind has forms . . . But on the other hand, in so far as this will result in speaking many times about the same affection because it belongs in common to many things, in this respect speaking separately about each one is somewhat silly and tedious. (644a29–34)

Mankind is Aristotle's standard example of an animal that does not fit into any of the wider kinds – it is, uniquely, a live-bearing biped. There is, of course, a closely related *aporia* in *APo.* 1.24, about whether "particular" demonstration or "universal" demonstration is better. The drive toward *partial* demonstration comes from the premise that what we want to understand is *ousia*, and that is surely to be understood by focusing on indivisible forms, not universals. The drive toward *universal* demonstration comes from the premise that to know that an attribute belongs to something, one must know it through that in virtue of which it belongs. To borrow an example from *PA* 1.4: to know that feathered wings belong to cranes or sparrows *not* in virtue of being cranes or sparrows but *in virtue of being birds* is to understand *better* why feathered wings belong to cranes and sparrows. Moreover, *APo.* 1.24 also stresses the value of understanding common attributes "in virtue of something common," precisely the language used in framing the "universal" option in *PA* I.1.

It is noteworthy that the issue is now worded as an *aporia*, and in the language of kinds and indivisible forms. This is the language of the two chapters on division, and the language used in the framing of the solution.

> Perhaps, then, the right course is this. In some cases – whenever kinds are spoken of by people in a clearly defined manner and have both a single common nature and forms in them not too distant – we should speak in common according to kinds, like bird and fish and any other there may be that, though it is unnamed, embraces, as a kind, the forms within it. But whenever they are not such as this, we should speak one by one, for example about mankind and any other such kind. (644b1–7)

Chapter 4 began by endorsing a way of identifying kinds with a common nature embracing closely related forms – namely, by identifying groups with multiple differentiae varying only by "the more and the less." So, having resolved this *aporia* by distinguishing cases when "speaking in common according to kind" is appropriate, he can conclude the discussion by invoking just that machinery.

> For their parts differ not by analogous likeness, as bone in mankind is related to fish-spine in fish, but rather by bodily affections, for example by large/small, soft/hard, smooth/rough, and the like – speaking generally, by the more and less. (644b11–14)

None of this machinery was available at the outset of the discussion, but following the introduction, in chapter 3, of multiple difference divisions moving from kinds with shared "universal differences" to indivisible forms with fully determinate differences, all that was needed was a way of distinguishing *between*-kind from *within*-kind differences. The introduction of the distinction between analogical and more/less differences in chapter 4 allows for Aristotle's final resolution.

Having traced the path to the solution to our first question and the framing of methodological principle about when and how to investigate what is "common according to kind," let us return to the first chapter and to its second question. It is a question about whether the *phusikos*, in studying animals, should follow a path traveled successfully by the astronomer.

> whether, just as the mathematicians explain the phenomena in the case of astronomy, so the natural scientist too, having first studied the phenomena regarding the animals and the parts of each, should then state the reason why and the causes, or whether he should proceed in some other way. (639b8–11)

Now it is well known that this is the way Aristotle thinks that mathematical astronomy has progressed, perhaps the most commonly cited text on the subject being the following passage in *APr.* 1.30:

> Thus the principles are provided by experience in each case. I mean, for example, astronomical experience provides the principles of astronomical knowledge; for when the appearances had been grasped sufficiently, astronomical demonstrations were easily discovered. And it is likewise with any other art or science. So that if the predicates about each thing have been grasped, we will be well prepared to exhibit their demonstrations. (46a20–4)

But while it may be "likewise with any other art or science," the above passage does not imply that the *way* one proceeds from observational experience to the principles will be the same in every case – nor that the principles arrived at will be the same. We are, after all, discussing a special class of natural substances that have a vast range of diverse appearances to be grasped and that come to be by means of temporally extended, complex, and coordinated processes. Thus, while the way of proceeding from phenomena to causes in this passage may be relatively unproblematic as a claim about mathematical astronomy, it is genuinely unclear what the implications of the general claim are for the investigation of animals.

It is thus no mistake, but a critical part of the narrative, that, rather than answering the second question, Aristotle introduces another, *third* question, that arises from "what we see concerning *natural generation.*"

> since we see more than one cause of natural generation, for example both the cause for the sake of which and the cause whence comes the origin of motion, we need also to determine, about these causes, which is naturally first and which second. (639b11–14)

In the case of mathematical astronomy the issue of prioritizing between two of the causes operative in nature will not come up, but it does in the study of animal generation. At the very least, then, the question of whether one should first study the phenomena and then their causes requires elaboration: since there are a number of causes involved in accounting for naturally generated beings, *which* cause should be taken up *first* after a study of the phenomena? And might there be some phenomena that are only to be studied after one of the causes has been?

There is, then, an interesting parallel between the two unresolved questions: both are initially stated as stark alternatives. The first requires the systematic development of a more complex theory of division before it can be answered, and it is focused on grasping the natures of animals as fully developed beings. The second concerns causal understanding and requires the more complex account of causation and necessity that Aristotle develops for the science of nature, including a view about the priority of goal causation to material and motive causation. The fact that the objects of central concern in *PA* 1.1, animals, are subject to unqualified, substantial coming to be has not, until now, been mentioned; and the second question cannot be adequately addressed until one takes seriously the differences between studying eternal, unchanging substances and substances subject to natural generation and destruction.

The argument at 639b15–21 defends the priority of teleological causation to the motive cause in "things composed by nature and by art" and ends by asserting, though not arguing, that "the cause for the sake of which" and the "beautiful" (*to kalon*)[17] are present *more* in the works of nature than in those of art. Following directly after that, in 639b21–640a1, Aristotle argues that necessity is not present alike in all natural things, and he notes a tendency among those who investigate nature to refer all explanations to necessity without taking account of this fact. He then develops this idea by introducing and defending a distinction between unqualified necessity and *conditional* necessity, the former present in eternal things, the latter present in all things that partake in generation. There is a link between the introduction of this distinction and the preceding argument for the priority of the final cause that is not always noticed. Aristotle's argument for the causal priority of that for the sake of which depends on the claim that in giving the account of a thing that comes to be, a doctor or builder will specify the state of health or the form of the house and will account for medical or building practice by reference to the goals they aim to achieve. He then uses the teleological nature of house building in explicating conditional necessity:

> It is necessary that a certain sort of matter be present if there is to be a house or any other end, and this must come to be and be changed first, then that, and so on continuously up to the end and that for the sake of which each comes to be and is. And it is the same way too with things that come to be by nature. (639b26–30)

The final section of this passage again builds on the two previous conclusions: 640a1–9 argues for a distinct manner of demonstration for contexts in which conditional necessity is operative, and where the starting points

[17] How to render *to kalon* in contexts outside ethics is a difficult question. Given that its primary reference is to beauty in form, I now think "the beautiful" is acceptable here, especially in light of 645a4–36. The message of that passage is that even animals that might appear disgusting to us are wonderful and beautiful once we focus on their causes and substantial being.

and definitions identify goals toward which change proceeds, provided nothing interferes.

> But the manner (*tropos*) of demonstration and of necessity are different in the natural and the theoretical sciences. (639b30–640a2)

There has long been debate about how to understand this remark, a summary of which is provided in Lennox 2001b, 128–31. One plausible reading of it fits well with the claim I am making about *PA* 1: that it is intended to take up the question, raised in *Mete.* 1.1, about the proper way to proceed with study of animals and plants. This remark about a different manner of demonstration and necessity suggests that the study of animals and plants is not to be pursued according to the methods appropriate for, say, the heavenly bodies, for it follows immediately after Aristotle introduces the distinction between *eternal* natural beings, governed by unqualified necessity alone, and *generated* natural beings, where definitions specify goals and where conditional necessity specifies the relationship of matter to such goals. This distinction is stressed in the justly well-known summative chapter of *PA* 1.5, which opens by contrasting the *eternal* natural substances with those that come-to-be and pass away, twice acknowledging that, while the former are more divine, the latter "take the prize with respect to scientific knowledge" (645a1–4; cf. 644b28–31). He then says:

> Since we have completed stating the way things appear to us about the divine things, it remains to speak about animal nature (*peri tēs zōikēs phuseōs*). (645a4–6)

Aristotle is, then, acutely aware of a fundamental demarcation in the natural world that brings with it the question of whether the study of animal nature is simply another part of the same *methodos* as the study of the eternal natural substances or whether the differences are sufficient to consider it a distinct investigation with quite different methods of investigation and modes of demonstration.[18]

Once all the machinery of teleology integrated with conditional necessity is in place, and a sketch of what demonstration in this context will look like has been outlined, Aristotle returns to the second of his unanswered questions and begins to develop an answer, at 640a9–14. There is not space here to explore the answer fully, but I want to show how an otherwise curious feature of his answer is illuminated by realizing that the answer grows naturally out of the development of the narrative to this point.

[18] This is a central theme of Falcon 2005 (cf. ch. 1, and ch. 4, 100–1). Falcon also has an interesting discussion (88–9) of the way in which the categorization of substance in *Metaph.* Λ serves as the metaphysical framework for this fundamental division of *natural* substance.

The passage begins as if it will provide a positive response to the question about whether one is first to study the observable phenomena about animals before attempting to search for their causal explanations – that is, as if it were an answer to question two. But it appears to morph rather quickly into a slightly different question: Should one try to understand the *being* of an animal as the necessary end result of its coming to be, or rather try to understand its coming to be as conditionally necessary for the being toward which it proceeds? Of course, he endorses the latter position – in those very words echoed in the passage from *Generation of Animals* 5 with which we began, and which in turn echo the *Philebus*.

For even with house building, it is rather that these things happen because the form of the house is such as it is, than that the house is such as it is because it comes to be in this way. For generation is for the sake of being; being is not for the sake of generation. (640a15–19)

On the account of *PA* I that I have been developing, these two questions are more intimately related than they might first appear to be. For suppose the "phenomena" you are concerned to study include both the anatomy and physiology – the parts and activities – of fully developed animals *and* the coming-to-be of those parts and activities. And suppose the causes include both the antecedent materials and moving causes out of which and by which development proceeds *and* the form and goal at the end of the process – that is, the functioning parts of full developed animals. The apparently uncontroversial suggestion that one should study the phenomena first, and then their causes, on grounds that knowledge of causal principles arises out of experience of the subject matter, now becomes quite complicated. For the processes and materials involved in an animal's coming-to-be can, from one perspective, be viewed as phenomena to be explained teleologically; but from another perspective can be viewed as the productive causes of the animal that comes to be.

I conclude this discussion of the narrative structure of the first four chapters of *On the Parts of Animals* by drawing attention to the gradual introduction of soul into the discussion. That animals must be understood in terms of their functional capacities is first introduced independently, partly by pointing out the limitations of thinking of them simply in structural terms, as an atomist might; and partly through reflections on the fact that artifacts are defined by reference to their functions (640b29–641a4). Even then, the argument moves cautiously. The concept of "form" is used only once in this discussion, and that use is with reference not to organisms but to beds (641a17). The word seems to be self-consciously avoided in the discussion of animals. That argument progresses, in outline, through the following steps.

1. Clearly one ought to state that the animal is a thing of a certain sort, and that entails identifying what each of its parts is (641a15–16).
2. Suppose that entails speaking either about soul, or a part of soul, or at least about what is ensouled (641a17–18).
3. [Recalling what we said contra Democritus about corpses vs. living bodies] without soul, at least, neither the animal, nor the part, is what it is except in configuration (641a19–20).
4. If the supposition in step 2 is accepted, then the *phusikos* must, at the very least, speak about that part of soul in virtue of which the animal is the sort of animal it is (641a21–3).
5. The *phusikos*, then, must say what it is – identify its *ousia* – and the attributes it has in virtue of the sort of *ousia* it is (641a23–4).
6. He must do that, given that a thing's nature is both its matter and its *ousia* (641a25).
7. The latter nature, its *ousia*, is also its nature both as mover and as *telos*.[19] (641a26).
8. It is the soul, or at least some part of it, that is this *ousia* (641a27–8).

It is a paper in itself to make the many interpretative decisions that need to be made to derive a coherent argument out of this. For now I only want to stress how different this is from the opening of *On the Soul* 2, where so much of the machinery of the central books of the *Metaphysics* is taken for granted and confidently mobilized. Here soul is introduced as a supposition, but even that supposition is weakened – maybe the *phusikos* only needs to talk about a *part* of soul; or perhaps all he needs to do is take for granted that what he is talking about is *ensouled*! It is a very long and complex argument just to bring us to the point where we are convinced that *if* the *phusikos* is aiming to grasp the nature of animals, *then* he is going to have to take seriously that animals are not just bodies configured in a certain way. Moreover, we are told next to nothing up to this point about what those functions are in virtue of which animals are the sorts of animals they are. This is an argument to the conclusion that the natural scientist must investigate soul, or animals *qua* ensouled, as part of his mandate; but the only content that is given to this claim is that this will involve investigation of functions and ends.

19 This is quite similar to the argument of *de An.* 2.4 that the soul is the cause of the living thing as form, moving cause, and that for which. What is striking, however, is that there is no reference to that discussion, and that οὐσία rather than form is used here.

Once again, then, in tracing the thread of the argument that ends up defending the idea that the natural scientist will need to study ensouled beings, *qua* ensouled, and that articulates what such a study will entail, *PA* 1.1 has a narrative structure, gradually articulating a positive program of research out of considerations of the shortcomings of previous natural philosophers. And it is fair to say that though chapter 4 concludes as if a discussion of the norms for investigating nature has been completed, what is still missing at that point is the fabric that the various threads are to yield when they are all woven together. That is only accomplished in chapter 5.

CONCLUDING THOUGHTS

I will conclude, then, by returning to chapter 5 and making some suggestions about viewing it as a unity. I myself have referred to the first half of it, 644b23–645b1, as an "intrusion."[20] It has no particle linking it to anything before. Moreover, this seems appropriate, since the preceding lines, the conclusion to our chapter 4, recounts what has been previously accomplished and apparently looks forward:

> So then: we have said how one ought to appraise the inquiry concerning nature (*tēn peri phuseōs methodon*) and in what manner the study (*hē theōria*) of these things might proceed methodically and with greatest ease (*hodōi kai hraista*); and further, concerning division, in what way it is possible by pursuing it to grasp things in a useful manner, and why dichotomy is in a way impossible and in a way vacuous. Having determined these things, let us speak about what comes next, making the following our starting point (*archēn tēnde poiēsamenoi*). (644b15–21)[21]

This is an accurate, if not terribly detailed, description of what has preceded. But the forward-looking reference is utterly lacking in content and, as previously noted, the opening lines of chapter 5 certainly do not connect with it.

Nevertheless *someone* put these beautifully crafted lines where they are, and I am going to entertain the thought that it was done for good reason. If one attends more to the content than to the style that has drawn so much praise, we see many connections with what has preceded. It begins with a contrast between two sorts of beings constituted by nature, eternal, ungenerated ones and those that partake of generation and perishing. These two classes of natural beings are contrasted both ontologically and

[20] Lennox 2001b, 172, 174.

[21] It should be noted that the terms "inquiry" and "study" are carefully chosen to echo the very first words of the treatise: *Peri pasan theōrian te kai methodon . . .*

epistemologically, and it is the plants and animals around us that win out epistemologically. But who cares how much we can learn about them if what we learn is demeaning and disgusting? Aristotle thus argues in elegant terms that examining the internal organs of animals is only disgusting if you forget to focus on the nature that crafted these things, that is, on their causes. If you do that, as we have seen, you come to realize that the goals for the sake of which things come to be and are, hold the place of the noble or beautiful in this investigation. And this turns out to be true whether the animal under investigation is itself deemed lowly or noble. You just need to keep focused on the nature and substantial being of the animal, and to view the investigation of its parts in that light. And that is to provide a vivid statement of where the argument of chapter 1 ends up.

But how, exactly, do we go about doing this so as to achieve the knowledge that hopefully you are now convinced is worth pursuing? Well: "it is necessary first to divide the attributes pertaining to each kind which are present in all the animals [of that kind] in virtue of themselves, and after that try to divide their causes." That is, we are now prepared to consider the integration of division of parts and division of activities or functions that is outlined in the second half of chapter 5, and which I discussed on pp. 64–5 above.

Here, then is my concluding conjecture about this chapter. As a whole, it can stand alone, intended as both an inspired defense of the hard-won conclusions of chapter 1, and a sweeping integration of those results with his discussion of the grasping of the differentiae of kinds and their forms at the appropriate level for explanation, and the means of identifying the kinds presupposed by that practice, that one finds in chapters 2–4. But though freestanding in that sense, it is also a fitting summation of the work done in chapters 1–4.

We end up, then, with the following complex unity of this apparent philosophical prolegomenon to the study of animals. The second half of chapter 5 is self-consciously an integration of the results of chapter 1 with the results of the arguments in chapters 2–4. Chapter 4 uses the new method of division achieved by the end of chapter 3 as a spring board for returning to the very first question discussed in chapter 1, namely, the question of the proper level of generality or specificity at which to begin an investigation of animals. The argument of chapter 1, in turn, is narrative in form, each question or topic growing out of, and presupposing, earlier results. Chapters 1–4, then, are far more tightly integrated than at first appears. Yet something is missing. There are a number of narrative threads that develop somewhat independently, beginning from different questions about how the investigation of nature should proceed; but those threads are not woven together.

Chapter 5, on the other hand, though probably not originally written as a continuous part of the methodological discussion of chapters 1–4, serves the function of weaving a whole fabric out of those threads. It may originally have been a free-standing unit – but only with it in place are you provided with a clearly integrated picture of how the *methodos* of animals is to be carried forward.

CHAPTER 4

An Aristotelian puzzle about definition: Metaphysics *Z.12*

Alan Code

The aim of this paper is to explore and elaborate a puzzle about definition that Aristotle raises in a variety of forms in *APo.* 2.6, in *Metaph.* Z.12, and again in *Metaph.* H.6.[1] My focus will be on the puzzle itself and its relevance to an inquiry into substance, but I will also touch on some issues pertaining to his proposed solution (or solutions) to it. One version of the puzzle is formulated in the aporetic section of *Posterior Analytics* as follows:

> T1: The same puzzle arises both for somebody proving by division and for this [above mentioned] manner of syllogism: why will man be a two-footed footed animal but not animal and footed? For on the basis of these assumptions there is no necessity that what is predicated becomes a unity, but it might be as if the same man were musical and literate.[2]

Here he is considering accounts that purport to define a thing by mentioning both a genus to which it belongs and one or more differentiae. The object of definition is not a word or expression such as "man," but it is rather some entity picked out or signified by that expression. What is at stake is the unity of the thing, that entity. Accounts constructed out of genera and differentiae may be arrived at as the end results of a method of division, and he further indicates that they may be arrived at as conclusions of syllogisms. Even if in the end this latter option is in some sense acceptable, regardless of whether one produces such definitions by a method of division or by deducing them from premises that include some other kind of premises,[3] we are told that the same problem arises for definitions of this sort.

What exactly is the problem? Suppose that the subject of the definition is man, and that the linguistic predicate of such a definition is, for instance,

[1] See also *Int.* 5.17a13–15. See Mary Louise Gill (Chapter 5, this volume) for an alternative interpretation of Z.12, and a novel treatment of H.6 and its relation to Z.12.

[2] *APo.* 2.6.92a27–33.

[3] Perhaps the premises even include some other kind of definition of the thing, on which see n. 47.

"two-footed, footed animal." What he asks is why man is a two-footed, footed animal, but man is not both animal *and* footed. If there is a problem here, then there should be something problematic about the unity of an item described not only as a man but also as both "animal *and* footed." For present purposes let us set aside the fact that Aristotle had earlier mentioned two differentiae (two-footed and footed), and concentrate on just this phrase.[4] Now of course if man is a footed animal it does follow that man is an animal and it does follow that man is footed. However, Aristotle seems to think that the methods under consideration provide no guarantee that the predicate in question constitutes a unity. Although one could reasonably urge that it is a unity in some sense, even so there is no guarantee that the predicate indicates any more unity than is captured by an expression such as "literate and musical." Accordingly, we should consider in what sense that would constitute a unity, and why unity in that sense is not sufficient for the predicate of a defining account.

Suppose, for instance, that the same man were both musical and literate. In that case one could correctly apply to that man both the predicate "musical" and the predicate "literate." The man is both musical and literate.[5] However, to say this of a man is not to predicate just *one* thing of him. Aristotle is not here offering an account of conjunctive predication, but if what is linguistically predicated is the expression "literate and musical," then the truth of such a predication involves two ontological predications of two different predicables. Insofar as the two predicables themselves indicate some kind of unity, this is due simply to the fact that they both apply to a single subject. It is not due to any intrinsic relation between these two predicables. There is a unity when there is some one other thing (i.e. a man) that happens to be characterized by both. That unity (the unity indicated when we say of a man that he is both musical and literate) is for Aristotle accidental. It might be the case that a man is both musical and literate, but the expression "musical and literate" does not indicate a single predicable, and the musical thing and the literate thing are one merely accidentally. The complex phrase is applicable to a man just in case the man is the subject for each of two distinct predicables. That being

[4] I am following the manuscript reading *zōion kai pezon* at 92a30. If Ross 1949 is right that the proper reading is *zōion kai pezon kai dipoun* then the conjunctive account would contain not two, but three elements. In that case the question of their unity would not be strictly parallel to that of musical and literate in 90a32–3. We shall see later that *Metaph.* Z.12 will offer an account of the unity of the two differentiae, footed and two-footed, according to which being two-footed entails being footed in that the former is a specific, determinate way of being footed.

[5] The predicate in this case would be the complex, conjunctive expression "musical and literate."

the case, the expression "musical and literate" does not signify one thing but picks out two different items, each of which the man (accidentally) is.

By way of contrast, if Aristotle thinks that there is a problem concerning the definition of man, then although it is true to say of a man that he is footed and it is true to say of a man that he is an animal, the predicate "footed animal" must be unified in a stronger manner, and a footed animal cannot be merely an accidental unity. The unity indicated by the predicate should not consist merely in the fact that each of the two constituent predicables applies to one and the same subject in such a way that that subject (the man) happens to be footed and happens to be an animal. The footed thing and the animal are not accidentally one, and being footed is not accidentally connected with being an animal through both simply happening to characterize a single, unified subject. What unites musicality and literacy is something distinct from both that serves as their subject, and the musical thing and the literate thing are each accidentally one with that subject. However, Aristotle takes it that it would be problematic if a footed animal were like this. Hence it would be problematic to hold that the same pattern of analysis is what unites the constituent elements of the phrase "footed animal." In particular, when saying that man is a footed animal one is not predicating two things of a man in the same way that an expression such as "musical and literate" does involve two distinct predications. The man is not accidentally one with either the footed thing or the animal, and they (the footed thing and the animal) are not one merely accidentally.[6]

Of course what one says about something when it is called footed is not the same as what is said about it when it is called an animal. Nonetheless, calling a man a footed animal is not saying two distinct things about it in the way that calling a man "musical and literate" is, and the subject that is called a footed animal is not related in merely accidental ways to what it is called when it is called a footed animal. In the *Posterior Analytics* passage Aristotle does not go on to elaborate on why it would be problematic if the two cases were instances of the same kind of unity, or if one's method of definition did not assure a relevant asymmetry. However, the elucidation of the puzzle in *Metaph.* Z.12 helps to make it clearer what Aristotle thinks is at stake. The kind of unity that a definition must have does not require that the items signified by the various components of the definition are identical, or that they are exactly the same predicable. However, it is required that

[6] As will emerge shortly, the musical thing and the literate thing are accidental beings that are accidentally the same as the man; musicality and literacy are attributes.

the definition indicates a kind of intrinsic or *per se* unity that contrasts with mere accidental unity. The chapter begins as follows:

T2: Now we should first[7] discuss definition, in so far as we have not discussed it in the *Analytics*; for the problem stated there is useful for our discussions concerning substance. I mean this problem: why is that thing, of which we say the formula is a definition, one, as for instance two-footed animal is of man; for let this be its formula. Why, then, is this one, and not many, animal and two-footed?[8]

As we have already seen, a version of this kind of puzzle was in fact discussed in *APo.* 2.6. We are now being told that there is more to say, and that it is worth discussing the puzzle further here because such discussion will be useful for our treatment of *substance*. In *Metaphysics* Z, unlike *Posterior Analytics*, Aristotle is concerned to answer the question "What is substance?," and now he indicates that pressing this puzzle is going to turn out to be useful for that project.

At least part of the reason why it is useful for the investigation of substance is that according to *Metaph.* Z.3 the essence of a thing has been thought to be the substance, or being, of that thing.[9] Since a definition signifies an essence, this would yield the result that the definition is the account of a thing's substance – it is the *logos tēs ousias*. Suppose that the account of man is reached by the method of division. If this account is in fact a *definition* of the nature of man, then it signifies the essence of man. As such it would be the account that expresses the *ousia*, or being, of a man (on the assumption that the essence of a thing is its substance). However, a definition signifies what something is intrinsically, in its own right. There must be an intrinsic, as opposed to an accidental, relation between a thing and its essence. As Aristotle put it earlier in *Metaph.* Z.4:

T3: And first let us say something about it [essence] logically. The essence of each thing is what it is said to be in virtue of itself.[10]

This is a perfectly respectable notion of essence, and one that he is here taking seriously as a candidate for the *ousia*, or being, of a thing. However, later I will give some indications that Aristotle rejects the idea that an account of essence reached through division is one that captures the *ousia*

[7] Picking up *skepteon husteron* at 1037a20. This is a first treatment of the aporia in the *Metaphysics*; H.6 will contain the second one.

[8] 1037b8–14. [9] 1028b33–4.

[10] *Metaph.* Z.4.1029b13–14; see also *APo.* 1.4.73a34–7 and *Metaph.* Δ.18.1022a25–9. I here use "logically" for *logikōs* instead of the revised Oxford translation's "in the abstract." For an illuminating discussion of this term see Burnyeat 2001, 19–25. I would urge that Aristotle is here discussing a notion of essence that is employed in dialectic of the sort discussed in his *Topics*.

or being of that item. Furthermore, the puzzle with which we are presently concerned seems relevant to this rejection because defining in this way is said to make the puzzle intractable.[11] My concern now, though, is to get somewhat clearer on the nature of the problem itself rather than to examine the reasons for Aristotle's rejection of this approach to capturing the substance of a thing.

In order to understand how this puzzle is relevant to a critical investigation of this approach to definition as a viable method of identifying the substantial nature of a thing, it is useful to look at some of Aristotle's remarks about how his predecessors inquired into substance. At a very general level, a method of definition by division (whatever its other uses might be) allows one to analyze a definable item into component parts or elements. If the item in question is a substance, this kind of analysis into parts yields a putative account of what that substance is. There was, though, a more "ancient" method that predated definition by division and that Aristotle sees as providing a competing model for producing accounts of the constituent elements of substances. This was a method employed by early natural philosophers that Aristotle represents as inquiring into the principles of substances by seeking those natural bodies that are the elementary constituents of which substances are ultimately composed. Aristotle construes them as searching for an account of a thing's substance (a *logos tēs ousias*, or account saying of that substance what it is) by attempting to enumerate the material constituents of which a thing is composed. He says of these "ancient" philosophers:

T4: And the ancients also testify to this by their work;[12] for it was of substance that they sought the principles and elements and causes. Well then, those today instead hold universals to be substances (for genera are universals, and on account of the logical nature of their investigation they say it is rather these that are principles and substances); but the old thinkers hold particulars to be substances, such as fire and earth, but not what is common to both, body.[13]

These old thinkers are early natural philosophers who searched for the substance of a thing by inquiring into its fundamental material components,[14]

[11] See H.6.1045a20–2. [12] I.e. that substance is the subject of this inquiry.

[13] *Metaph.* Λ.1.1069a25–30, replacing "in the abstract" with "logically," for which see n. 10.

[14] He elsewhere will say such things about them as "Now the ancients who first began philosophizing about nature were examining the material origin and that sort of cause" (*PA* 1.1.640b4–6). Translations of *On the Parts of Animals* are from Lennox 2001a. See also *Ph.* 2.2.194a18–21: "Looking at the ancients, it [natural science] would seem to be about the matter, for Empedocles and Democritus touched on form and essence only to a small extent." As T4 indicates, he sees these early students of nature as attempting not just an inquiry into the natural phenomena of the physical universe, but

and they are contrasted with those "today" who as a result of their logical method of examination treat universals as substances. *Metaphysics* B discusses these two allegedly rival approaches to the same question, and one of its puzzles concerns whether it is those earlier ancient natural philosophers or the modern logic-choppers who have the correct view about the method to use in searching for an account of the substance of a thing. The relevant puzzle begins as follows:

T5: There is great difficulty as to how one should put things so as to attain the truth, both concerning these things[15] and concerning the principles, whether one ought to take the genera as elements or principles, or instead the primary constituents present in each thing.[16]

Corporeal, physical elements have a claim to be basic and fundamental. However, opposed to the views of the ancient natural philosophers is a very different approach that employs a very different kind of method. According to this more recent method an account of the being or *ousia* of a thing is constructed out of genera, and these items are its elements and principles.[17] On this view the genus is a principle, and a principle of such a sort as to be described as an *element*. Both camps employ methods for investigating those principles that they take to be the elements of the things that are, but they disagree whether the proper method is to look for those elements that are material constituents or those that are generic kinds. As Aristotle points out, however, they cannot both be right:

T6: But it is not possible to speak of the principles in both ways. For the formula of the substance is one; but definition through genera will be different from that which states the constituents.[18]

as also engaged in an inquiry into *substance*. This very same enterprise – inquiry into substance – is engaged in by others as well, including not only those contemporaries who argue that the genus is substance, but also Aristotle himself in *Metaphysics* Z. However, even for the purposes of natural philosophy their method is inadequate. Natural philosophy must, he thinks, inquire into matter, but it must do more. As he says in *PA* 1.1: "It is clear, then, that these natural philosophers speak incorrectly. Clearly, one should state that the animal is of such a kind, noting about each of its parts what it is and what sort of thing it is, just as one speaks of the form of the bed" (*PA* 1.1.641a14–17). They failed to give a correct account of the substantial nature of a physical object precisely because they did not have a method for investigating and defining forms, and it is in this regard that the more "contemporary" philosophers represent significant improvement. As he puts it: "One reason our predecessors did not arrive at this way is that there was no 'what it is to be' and 'defining substantial being'" (*PA* 1.1.642a24–6).

15 This refers to the preceding, fifth puzzle of *Metaphysics* B. It begins at 997a34–5 by asking "must one say that there are only sensible substances, or that there are others besides these?"

16 *Metaph.* B.3.998a20–3.

17 Aristotle is here using the concept of genera in a fairly broad way so as to encompass both the genus and the differentia.

18 *Metaph.* B.3.998b11–14.

The disagreement about principles that he is describing is one that he takes to be a dispute about what the account of the substance of a thing is, and what method one ought to employ in seeking such an account. Since he construes the issue in this way, he represents them as having divergent views both about substance and about methodology in metaphysics. These divergent approaches of his predecessors are part of the larger background to the inquiry into substance that he initiates in *Metaphysics* Z.[19]

Metaph. Z.12 takes up the question of the unity of a definable entity in connection with one of these two methods, the one that attempts to establish a definition by a process of division. If we view this in the context of the kind of dispute just mentioned, it is clear that the items articulated in the kind of definition of a substance he is examining are elements in the broad sense of genera rather than material constituents. Furthermore, as we shall see, his solution in that chapter involves presenting a determinate view about the account of the substance of a thing. So let us now resume our examination of that chapter and the way it elaborates the puzzle:

T7: For whereas in the case of man and pale they are many when one does not belong to the other, but they are one when one does belong and what underlies, man, is modified somehow (for then they become one thing and the pale man exists), in the present case one thing does not participate in the other, for the genus is thought not to participate in its differentiae (for the same thing would participate in opposites simultaneously, for the differentiae are opposites by virtue of which the genus is differentiated).[20]

In T2 he made it clear that the puzzle concerns definable items, and in each of the passages the definable item in question is *man*. The kind of definition of man put forward is one that mentions both the genus animal and the differentia two-footed. Our initial text from *Posterior Analytics*, T1, mentions an additional differentia, footed. As we shall see, this additional differentia makes its appearance later in Z.12 in that chapter's solution to the puzzle.

For Aristotle genera and differentiae are beings, or things that are. When an item such as man is defined by genus and differentia the definition is a complex phrase that contains parts that signify these beings. The terms "animal" and "two-footed" occur in the definition, and the former signifies a genus, and the later signifies a differentia. These are both beings, and since the differentia two-footed distinguishes the species man from other

[19] In that work he does not resolve this puzzle by siding with either the early natural philosophers or those who employ the more modern approach. It is relevant to point out that in *Metaph.* Z.17 he argues for the view that the substance of a thing is not one of its elements.

[20] *Metaph.* Z.12.1037b14–21.

members of its genus they (the genus and the differentia) must in some sense be different beings. However, if the genus is one being and the differentia another, then the definition "two-footed animal" mentions two beings. Furthermore, if man is by definition a two-footed animal, and the relevant definition mentions two beings, then the question arises as to why the entity defined (i.e. man) is not a plurality that consists in both a two-footed thing *and* an animal that is distinct from that

As I have already noted, it is surely correct to say of a man that he is both two-footed and an animal, for the same thing can be both two-footed and an animal. However, the way in which this is the case must differ significantly from the way in which a musical thing and a literate thing can be one and the same, and that is what we now need to clarify. The *Posterior Analytics* passage elaborates the puzzle by suggesting that "it might be as if the same man were musical and literate," but that passage does not attempt to explain what the problem would be if things were like that for the definition of man. A definition reached by division *inevitably* mentions more than one item, and this will turn out to be problematic if the predicate in that definition indicates a unity merely in the same way that "musical and literate" does.

As already noted, being literate and being musical are one when one and the same man is both literate and musical. The very same man can be both literate and musical, but absent this connection with a common subject there is nothing that unifies these two beings. We need to look a bit more closely at the reason why this cannot be the way in which man is both two-footed and an animal *if* being a two-footed animal is the content of the definition of that subject (i.e. of man).

What needs to be added to the account of accidental unity already discussed above is the crucial role that *substances* play as the subjects that account for the unity in question. Literacy and musicality are attributes that may be possessed by certain *substances*, and if and when a substance is characterized by both of these simultaneously it is both a literate thing and a musical thing. One may truly say of some *substance* both that it is literate and that it is musical. What unifies this literate thing and this musical thing is the connection that each has with a single substance. What is here being highlighted – and was absent from my earlier discussion of accidental unity – is that for Aristotle the unifying *subject* alluded to in each case is a substance, whereas such items as the literate or the musical thing are not substances (although each is accidentally the same as some substance). The substance (e.g. Coriscus) could have existed without being either literate or musical and could continue to exist as the very same substance even were he to lose the attributes by virtue of which he is correctly called musical and literate. Alternatively, he could be musical, but not literate, or vice versa. Even so, the literate and the musical are one and the same thing when the substance, Coriscus, is characterized by both musicality and literacy.

Consequently, the reason why the literate thing and the musical thing are one and the same is simply that each is one and the same as the same particular *substance*. Consider, for instance, the discussion in *Metaph.* Δ.6 as to how Coriscus and the musical are accidentally one:

T8: . . . things are called one accidentally such as Coriscus and the musical, and musical Coriscus (for it is the same thing to speak of Coriscus and the musical, and musical Coriscus), and the musical and the just, and musical and just Coriscus. For all these are called one accidentally, the just and the musical because they are accidents of *one substance*, and the musical and Coriscus because one is accidental to the other.[21]

Coriscus is a substance, musicality is one of his accidental attributes, and the musical Coriscus is an accidental unity. Furthermore, the musical thing and the just thing are accidentally the same because each is accidentally the same as a single substance. In the account or formula "musical Coriscus" one part picks out *musical* and the other part picks out *Coriscus*. What makes musical Coriscus a unity is simply that the former part is an accident of, or coincides, in the latter.[22]

This kind of account of the unity of accidental beings makes it clear that the common unifying subject is a *substance*. Further, it presupposes that the substance itself is a unity in a way that is more fundamental than the accidental unity of a substance with an attribute. The just and the musical are one in that each is said to be an accident of one substance. Unlike the musical Coriscus or the just Coriscus, both of which are accidental unities, Coriscus himself is simply one substance. Coriscus is not an accidental unity in the sense that he is really a complex or compound of some other more fundamental substance that serves as a subject and one or more accidental attribute. By way of contrast, Coriscus is, for Aristotle, an intrinsic or *per se* unity. In general terms, while an accidental unity may be analyzed into an intrinsic unity and an accidental attribute, the intrinsic unities are not to be analyzed in that way.

[21] *Metaph.* Δ.6.1015b17–23.

[22] 1015b23–7. On Aristotle's behalf we should observe a distinction between accidental beings such as "the musical thing" and the attribute musicality. Coriscus is accidentally the same as the musical thing but is not accidentally the same as the property of musicality. Musicality is not an accidental being – rather it is one of the properties that may accidentally characterize a substance such as Coriscus. See Frede 1987 (esp. 41–2).

Given that what it is that unifies the literate and the musical thing (assuming that they are unified) is simply their relation to one and the same substance, there is for Aristotle nothing problematic about the claim that a man is both literate and musical. The literate and the musical constitute a unity, albeit a merely accidental unity. However, the account of unity that applies in this type of case could not help us to understand how a man could be both two-footed and an animal *if* a man is substance, and as such an intrinsic unity, and "two-footed animal" is the account saying what a man intrinsically is.[23]

In light of T8 we should think of asking why the musical Coriscus is one as asking why Coriscus and the musical are one. If the formula "biped animal" were parallel to this type of case, then asking why a footed animal is one would be the same as asking why the footed thing and the animal are one.[24] So let us now see what would be the case if the predicate "two-footed animal" were susceptible to the same pattern of analysis as Aristotle uses to explicate how a man could be both literate and musical. If it were, then the two-footed thing and the animal would be one and the same thing because each would be accidentally the same as a single substance, the man. The two-footed thing and the animal would be accidentally the same thing because each is accidentally the same as a single substance, the man. However, then the man would be accidentally two-footed and accidentally an animal, and so the formula "two-footed animal" would not express what a man is intrinsically, or in its own right.

A man, though, is a substance, and as such is an intrinsic unity, and an account saying what man is intrinsically cannot be the same as an account that says what man is accidentally. Since a definition says what a thing is in its own right, it would then follow that the account arrived at by collection and division would not even be a definition of man. The elements of the formula would be unified only in the way that literate and musical are unified through their connection to a single substance as the common subject that merely happens to possess them. Such a formula does not

[23] Aristotle has earlier claimed that in the strict and unqualified sense there is definition (and essence) only for *substances* (Z.4.1030b4–6). There are various derivative ways in which there are definitions for non-substances, but what we are concerned with in this *aporia* are definitions of substances. A substance is a being in its own right, or intrinsic being, and the definition of a substance says just what it is in its own right. However, if a definition does not say what some substance is in its own right, then it is not a definition in the strict sense and is not the definition of a substance. Furthermore, if the account saying what man is shows that a man is an accidental unity of some sort, then (i) that account is not a definition in the strict sense, and (ii) a man is not a substance.

[24] See *Metaph.* Δ.6.1015b18–19 for the musical Coriscus case.

capture what that subject is in and of itself, and hence it would not be the definition of that subject.

Alternatively, since the unity in question here cannot be due to a relation that both elements have to some underlying substance, one might consider treating "two-footed animal" as an analogue to "musical Coriscus." However, this too fails to yield a satisfactory account of the intrinsic unity of a substance. Musical Coriscus is an accidental unity of substance and attribute, and it is a unity because *musical* is an accident of the substance Coriscus. If we were to treat a two-footed animal as unified in this way, then either "two-footed" would be an attribute of an animal, or "animal" would be an attribute of something two-footed. In either case the formula would still be that of an accidental, not an intrinsic, unity. In the development of this puzzle in *Metaph.* Z.12 Aristotle considers the former type of approach, though not the latter. In the former case the genus "animal" would correspond to the underlying substance "Coriscus," and the differentia "two-footed" would correspond to *musical*. Hence, on the analysis in question the genus animal would be a subject as *substance* and the two-footed thing would be accidentally the same as that substance.[25] Given this pattern of analysis, the genus, being the subject and substance, would have to play the role of the intrinsic unity. Against this Z.12 introduces the idea that:

T9: for the genus is thought not to participate in its differentiae (for the same thing would participate in opposites, for the differentiae are opposites by virtue of which the genus is differentiated).[26]

Since he is here assuming that the relation between Coriscus and pallor is one of "participation," the reference to participation in this argument is arguably to some Platonic conception.[27] The point is that if one were to treat the genus as a substance and the differentia as the result of an attribute inhering in that substance, then the genus would have to possess opposing properties. For instance, if man is defined as "two-footed animal" and horse as "four-footed animal," and if both man and horse have the same genus, then a single substance (i.e. animal) would be equally two-footed and four-footed. However, a single substance cannot be both two-footed and four-footed. Consequently, the genus is not a substance that is accidentally the same as the two-footed thing.

[25] Additionally, the attribute of two-footedness would correspond to the attribute of musicality.

[26] *Metaph.* Z.12.1037b18–21.

[27] In any case, it would not be relying upon the definition of participation in the *Topics* according to which participation must be essential predication (*Top.* 4.1.121a11–13). Thanks to Marko Malink for this and the texts mentioned in the next note.

One might try running the analysis the other way by treating the differentia as the subject and the genus as the attribute. However, Aristotle elsewhere indicates a reason not to go by this route since:

T10: it is impossible for the species of a genus to be predicated of their own differentiae or for the genus to be predicated without its species.[28]

Accordingly, the differentia "two-footed" could not be a subject of which the genus "animal" is predicable since no species of that genus is predicable of two-footed, and the genus is not predicable of a subject if none of its species are.

In the subsequent lines in the Z.12 passage[29] Aristotle goes on to say that even if the genus does participate in the differentiae the same account applies if there is more than one differentia. That is, the problem of unity arises when there are multiple differentiae, even if we were to put aside the impossibility of a genus participating in the differentiae. For instance, for man the differentiae might be: footed, two-footed, and wingless.[30] The context allows for more than one interpretation of the exact formulation of the problem he has in mind. He could mean that the *problem* of unity cannot be solved even if we do allow that the genus is a subject for differentiae, since even if the differentiae all were unified in that they were accidental attributes of one substance, there are still many of them. Even if the genus were a subject for each of them, the definable object, man, would simply be an animal that happens to be a footed thing and happens to be a two-footed thing and happens to be a wingless thing. We would end up treating the object that we are trying to define (in this case, man) as a series or a plurality of accidental complexes and not as a single intrinsic unity. In that case it could not be a substance.

In fact, he could also have pointed out that we could get a similar result even if there were just a single differentia, and the definition had just two

28 *Metaph.* B.3.998b24–6. Granted, this occurs in the context of the development of an *aporia* and so we cannot automatically assume it represents his own view. However, even if we did not have a text showing that for him the genus is not predicated of the differentia, the main point remains intact. (But see *Top.* 6.6.144a32–b3; cf. also *APo.* 2.3.90b34–8. Also, the differentia cannot partake of the genus: *Top.* 4.2.122b20–3; cf. also *Metaph.* K.1.1059b33.) Regardless of which of the two elements in the account is taken to be subject and which predicate, if the relationship between them is accidental, then the subject they purport to define would itself be an accidental unity of some sort, and not a substance.

29 1037b21–7.

30 Notice that "wingless" is not a determinate way of being footed, nor is being footed a determinate way of being wingless. Additionally, in *PA* 1.3 he classifies it as a privation and denies that it divides into more determinate forms. We will see shortly that it is not the kind of differentia that fits into the solution to the problem of unity that Z.12 presents.

elements.[31] Even in this limiting case, man would not be a substance, but a kind of complex of substance and attribute, and the formula of man would be an account of an accidental compound, not the account of a substance. However, on the assumption that a man is a substance, this cannot be right since:

T11: But nonetheless as many things as are in the definition must be one; for the definition is some one formula and a formula of *substance*, so that it must be a formula of some one thing; for substance signifies some one thing and a "this," as we say.[32]

Nothing turns on the particular example, or even on whether the example is an acceptable exemplar of definition by genus and differentiae. For purposes of the general point it does not even matter whether this example chosen, in this case man, is a substance or not. This is a general puzzle for the view that an account arrived at by division is a definition of a substance, of an *ousia*.

This helps us see how the puzzle stated in *APo.* 2.6 is relevant to an inquiry into substance. It is not just that the predicate in a definition must signify a unity. It must signify an intrinsic unity as opposed to an accidental unity. This in turn is required because a definition says of some definable object what it is in its own right, intrinsically. However, the definable object itself would not be an intrinsic unity if the only account saying what it is has parts that are related in non-intrinsic and accidental ways. If the account saying what man is fails to express an intrinsic unity, and yet this is the account that gives the being of the item in question, then the object of the account, man, also fails to be an intrinsic unity. But in that case it cannot be a substance. The challenge is not to show that each of the parts of the definition somehow picks out one and the same identical thing, but

31 This being the case, one would want an explanation as to why Aristotle here brings up a plurality of differentiae. *Perhaps* it is simply because he thinks that definable species will always have more than one differentia. However, there also is a special problem that arises in the case of multiple differentiae. Since his ontology of accidental beings has not provided for conjunctive accidental beings such as "the literate and musical man," the analysis of accidental unity could not be applied to cases such as "the footed and two-footed animal." So even if we could skirt around the objection that a subject cannot be characterized by opposites, we would not be able to apply an analysis of accidental unity to the definition of man once multiple differentiae are brought in.

Another possibility worth entertaining is that he has in mind the maneuver alluded to in Z.14.1039a26–8 that takes the animal in man as different from the animal in horse. Such a move would escape the objection that one and the same entity (namely animal) is characterized by contraries. On such a proposal, the animal in man is arguably not characterized merely accidentally by two-footedness but rather *is* that animal that is intrinsically two-footed. This provides a potential escape route to the objection about opposites being present in the same subject, but the problem resurfaces once we take into consideration the plurality of differentiae.

32 *Metaph.* Z.12.1037b24–7.

rather to define the being of a substance in such a way that each of the parts constitutes an intrinsic unity. The definition is complex, but its parts must be related in such a way that collectively they constitute what a single *substance* is in and of itself.

This is a general problem for definitions, since all definitions are complex phrases that contain constituent parts that themselves signify beings. The solution to the problem, in general terms, requires addressing the following:

T12: And in the case of definitions, how the things in the formula are parts, and why the definition is a single formula (for it is evident that the thing is one, but in virtue of what is the thing one, given that it has parts?) must be considered later.[33]

The definition of an intrinsic being is linguistically complex, but what is signified by its various components must be intrinsically unified. The problem of the unity of a definable object requires giving an answer to the question "in virtue of what are the various parts unified?" that is different from an account of a merely accidental unity. Whatever the answer to this question, it cannot be that what unifies the parts is that they are accidental attributes of some subject, or that one of them is a subject of which the other parts are accidental attributes. If the account by genus and differentiae is the account of the substance or being of a definable, intrinsic being, then its elements cannot be unified by their relation to something else, something other than those very elements. Hence the account itself must contain reference to whatever it is in virtue of which the object of definition is one. But what is that?

There is an answer to this question presented in *Metaph.* Z.12, though it does not seem to be Aristotle's final word.[34] The solution is that what he calls the "final differentia" is the form and substance of a thing.[35] If, for instance, the final differentia for man is two-footed, then that single differentia, two-footed, is the form and substance of man. On this solution we dispense with the assumption that the entire definition by genus and differentia

[33] *Metaph.* Z.11.1037a17–20. The solution is first attempted in Z.12 but is given in a different manner in *Metaph.* H.6 (for which see Gill, Chapter 5 in this volume).

[34] As just stated in the previous note, when the puzzle about unity is raised once again in H.6 the resolution is different from the one given in Z.12. *Metaph.* H.6 invokes the distinction between matter and form, and the correlative distinction between potentiality and actuality. Whatever we make of the H.6 solution to the problem of unity, it does not presuppose the method of definition by division. In fact, prior to giving his solution there he indicates that the puzzle cannot be solved by those who define in that manner (1045a20–2). Furthermore, and most importantly, it does not invoke the idea that the final differentia entails the rest and as such is the substance of the thing defined.

[35] 1038a25–6.

gives the *ousia* of man and replace it with the idea that the account of the *ousia* of man is an account constructed solely out of differentiae.[36]

He argues for this in two main steps. The first step is to rule out the possibility that the genus is the unifying substance of the thing,[37] and the second is to argue that the multiple differentiae in the definition arrived at by division can be put in a linear order in such a way that when dividing the more general into the more specific, each differentia is a "differentia of a differentia." For instance, "two-footed" is a differentia that divides footed. Since the final and most determinate differentia entails all of the higher and more general differentiae, the account of the substance of man is reducible to a single term: "two-footed." The ultimate answer to the question "What is man?" would in that case simply be "two-footed." This answer has none of the internal complexity of a definition by genus and differentia. Since it does not mention a plurality of parts, the question as to what unifies those parts does not even arise. And if this is correct, then we have discovered *the* constituent element in a definition by division that is that by which the parts are one.

As noted, this solution assumes that definitions are arrived at by division, and further that the final differentia entails all of the others in that it is a more determinate and specific realization of those more general differentiae. However, this is at odds with what Aristotle says about the plurality of differentiae in the methodological discussion found in *PA* 1.3. That chapter sums up one of its main conclusions by saying that "it is impossible for those who divide in two to grasp any of the particular animals."[38] Although that chapter acknowledges that in some cases a plurality of differentiae are indeed unified in the way explicated in *Metaph.* Z.12,[39] it also insists that animals need to be defined by a group of

[36] 1038a28–30.

[37] For this purpose he invokes the claim that the genus does not exist in an unqualified way over and above its species, or if it does, it does so only in the way that matter does (see 1038a5–6). The former is partially supported by the earlier argument against treating the genus as a subject of which a differentia is an attribute when earlier Aristotle was considering the attempt to treat the genus as a substance and the differentia as an attribute, and showing that this leads to absurdity. Although this does not prove that the genus is not a substance, one obvious way to avoid that absurdity altogether would be to deny that the genus does have independent existence. The assumption would be fully supported by an argument that no universal is a substance, and this is in fact one of the main themes of the next chapter. On the latter option (genus as matter) see Gill (Chapter 5, this volume).

[38] *PA* 1.3.644a10–11.

[39] See, for instance, 643b32–644a1: "For the continuity of the differences derived from the kind according to its division means just this, that the whole is a single thing. But the mode of expression makes it seem that the final one alone is the difference, e.g. 'multi-split-footed' or 'two-footed,' and that 'footed' and 'many-footed' are superfluous." See also 1.2.642b5–9.

differentiae that do not all fall under a single, continuous division in this way.[40]

One might query whether *Metaph.* Z.12 was written before this methodological discussion in *On the Parts of Animals*, but even if it were, that does not alter the fact Z.12 is drawing out consequences for the process of arriving at definitions through a method of division. Whether or not Aristotle's considered view is that this approach can solve the problem of the unity of definition, it still does provide an argument that purports to identify that by which a substance is one *if* we search for the substance of a thing by examining the contents of definitions reached through that method. This is a method that needs to be taken into consideration in a search for the account of the substance of a thing, and Z.12 is a part of his examination of the use of that method in metaphysics.

Among other things, the conclusion that the substance of a thing is simply the final, and most determinate, differentia arrived at through division provides an answer to the following question posed in *Metaphysics* B about which of the elements in definitions by genus and differentia are the principles:

T13: Additionally, even if the genera are principles most of all, ought one to consider the first of the genera to be principles, or the last ones which are predicated of the individuals?[41]

According to Aristotle this was a matter of dispute. This statement immediately follows T6, and the context indicates that T13 is a further problem specifically for those who take the *logos tēs ousias* to be the account constructed out of genera (broadly construed). Granted that the genera broadly construed are principles, which element in a definition is the substance of the definable item? The argument in Z.12 would resolve this further dispute in favor of the ultimate differentia.[42] This complements nicely the puzzle in *Metaph.* B.3.998b14–28. There, after arguing in favor of the highest genera being the principles, and identifying those with being and unity, it then invokes T10 to construct an argument to show that it is impossible for these even to be genera. So, if we are going to search for the *ousia* of a thing by examining the component elements of a definition by genus and differentia, it would have to be the differentia that receives this honorific title, and the most specific, determinate one at that.

[40] 644a6–11; cf. 643b10–17 and 23–4.

[41] *Metaph.* B.3.998b14–16.

[42] It is worth noting that *Metaph.* H.2.1043a2–7 does show commitment to the view that differentiae are the cause of being for a substance. However, that passage does not endorse the Z.12 solution to the problem of unity and does not indicate that such differentiae are arrived at through a method of division.

However, what Aristotle says regarding definition by genus and differentia in the *Topics* would suggest that this is the wrong answer. Although "the genus and differentia constitute the formula of the species,"[43] the differentia is treated in the *Topics* as some essential qualitative determination of the thing, and as such not as *what* the thing is:

T14: . . . for a thing's differentia never signifies what it is, but rather some quality, as do footed and two-footed.[44]

Hence, of the elements in the definition, the element that is in the category of substance will be not the differentia but the other component, the generic kind. Hence following this method one is led to the view that it is the genus that is the *ousia*:

T15: . . . of the items in the definition, the genus is most of all thought to signify the substance of the thing defined.[45]

Thus in the *Topics* a formula or account of a species is constituted by the genus and the differentia, and of those elements, the genus is the *ousia* of the thing (whereas the differentia is a quality). However, the argument in *Metaph.* Z.12 turns this on its head with its argument that the other element, the differentia, must be the *ousia* of a thing.

Against this background I would suggest that its treatment of the puzzle about unity is intended to show that *if* one attempts to locate the *ousia* of a thing by singling out an element of a definition arrived at by dichotomous division, one is led not to the view that the genus is substance, but rather to the opposed view that the substance of a thing is a final or ultimate differentia that is specified in the definition reached by division. However, whatever reasons there are in favor of the view that such a differentia is a quality would also be reasons to reject this approach as leading to a correct view about substance. If so, in this chapter Aristotle is not endorsing division as the appropriate method for reaching a correct account of the substance of a thing. Accordingly, we need not read the chapter as an attempt to show how he himself would utilize that method successfully in defense of what he himself takes to be the substance of a thing. There is fairly widespread scholarly consensus that for Aristotle in the case of a

43 *Top.* 6.6.143b19–20.

44 *Top.* 4.2.122b16–17 (for purposes of consistency I use "footed" and "two-footed" for *pezon* and *dipoun*). There are quite a few other passages in the *Topics* that treat the differentia as a quality, including 4.6.128a26–7 ("also that the differentia always signifies a quality of the genus") and 6.6.144a20–2 ("See also whether the given differentia signifies a certain 'this' rather than some quality; for it is thought that every differentia shows a quality.").

45 *Top.* 6.1.139a29–31.

living thing its substance is its form, and that this is its soul. However, this is not a view that Aristotle arrives at through an examination of the parts of definitions arrived at by genus and differentia.

If this is right, then Z.12 is best seen as part of a critical examination of a method for inquiry into substance that fails, and this in turn helps show the need for a different mode of inquiry. *Metaph.* Z.17 initiates a fresh inquiry into substance by starting with the idea that a substance is a cause and a principle:

T16: Since, then, substance is a principle and a certain cause, one must pursue it from this standpoint. The "why" is always sought in this way: why does one thing belong to some other thing?[46]

It is here that Aristotle invokes ideas about the connection between syllogistic argument and definition that he develops at some length in *Posterior Analytics* 2.[47] Aristotle stresses in Z.17 that in posing this kind of question we are not asking why a man is a man. The answer to that is either something to the effect that each thing is itself, or perhaps that a man is indivisible with respect to itself (and that just is what being one is). In either case one would be giving an answer that could be applied to anything at all. Such an answer gives us nothing that is distinctive of man, and it does not require any real inquiry at all.[48] However, where there is a genuine inquiry the question must be posed in the form "why does something belong to some other thing?," and accordingly a subject and a predicate must be distinguished so that we may inquire into why something (i.e. the predicate) belongs to something else (i.e. the subject). The subsequent inquiry would be a search for an answer to the question of why that predicate belongs to that subject. When it is formulated in this way, Aristotle thinks that it is possible to ask why a man is what he is:

[46] *Metaph.* Z.17.1041a9–11.

[47] His discussion there is intricate and full of subtleties, and this is not the place to examine it. For present purposes it suffices to note that it distinguishes different kinds of definition: "Hence one definition is an indemonstrable formula of what a thing is; one is a deduction of what it is, differing in arrangement from a demonstration; and a third is a conclusion of the demonstration of what it is" (*APo.* 2.10.94a11–13). The first of these three types of definition is an indemonstrable first principle of a science. Being indemonstrable there simply is no middle term that mediates or explains the connection between its subject and predicate. If definition by genus and differentia were to fall into this first type, then there could be no inquiry as to why a man is a biped, or why a man is an animal. Conversely, if there are answers to questions of this type, then the definition by genus and differentia is not a definition of the first type, but rather could be (when properly explained) the *conclusion* of a demonstration in which the predicate in the indemonstrable account is used as the middle term.

[48] See *Metaph.* Z.17.1041a11–20.

T17: But one might well inquire why man is *such and such* an animal.[49]

The phrase "*such and such an animal*" (*zōion toiondi*) is a complex expression. There are all sorts of terms that we could insert at the placeholder "*such*," and one of them would be "two-footed." Among other things, one could engage in genuine inquiry as to why a man is (a) *two-footed* animal. For such an inquiry to be successful there would have to be a middle term that connects the other two terms in such a way that there is an explanation of why the predicate in question belongs to the subject, man. Such an explanation would have as its conclusion the proposition that man is a two-footed animal and would contain in its premises some third term that explains or mediates the connection between being a man and being a two-footed animal. From this perspective, the definition by division is not the account that captures the being or *ousia* of man. The definition by division may well be a necessarily true proposition that classifies the species by isolating a factor that distinguishes it from other members of its kind, but it does capture the *ousia* of the definable item.

Given that the soul is the substance of a living thing, it would be the possession of a soul of a certain type that explains why something is the kind of living thing it is. Animals, for instance, are *perceptive* living things. On this hylomorphic conception of the soul as substance, the various features by reference to which different kinds of animals are distinguished and classified are themselves explicable by reference to a single cause, the soul. The definition of man by division, however, is not foundationally bedrock. Nonetheless, if the methodological approach put forward in Z.17 does succeed, then it also suggests an alternative approach to solving the puzzle about definition presented in Z.12. As I have argued above, the problem there was to show how the different elements in the definition by division constitute an intrinsic, as opposed to an accidental, unity. If the definition by division can itself be explained by reference to a middle term, that thereby would show how and why the features in the definition by division are intrinsically, as opposed to accidentally, related to the defined substance (i.e. man). In that case, definitions by division would not answer the question posed in T12 by revealing that by virtue of which a definable item is one.[50] However, there would be hope that a definition of a different kind, reached by a different route, could.

[49] *Metaph.* Z.17.1041a20–1. Here I follow the Oxford translation in taking "man" as the subject.

[50] Hence the *ousia* of a thing is not even mentioned in such an account, for which see *Metaph.* H.3.1043b10–14.

CHAPTER 5

Unity of definition in Metaphysics *H.6 and Z.12*

Mary Louise Gill

In *Metaphysics* H.6 Aristotle asks: "What unifies a definition, making its parts into a whole and not a mere list?" A definition is a single account, he says, not by being tied together like the *Iliad* (the whole poem in twenty-four books), but by being of one thing. Although a definition is a complex formula, consisting of a number of linguistic parts, it is one if the object defined is one. So the unity of definition is parasitic on the unity of the object defined. Aristotle is looking for basic entities, whose being is fully determined by their essence and not by other entities which claim priority. These basic entities ground our understanding of other things. What, then, makes an object one, whose account is a definition? Aristotle discusses three sorts of examples: substantial forms, material composites, and categorial properties.[1]

Start with a determinate form such as man.[2] What makes man one? Suppose we define man as "biped animal": why is man one thing and not an aggregate of two things, animal and biped, which are distinct from each other and logically prior to it – that is, mentioned in its

It gives me great pleasure to dedicate this chapter to Allan Gotthelf. Its topics relate to his own pioneering research on division, definition, and the unity of substances. My interest in these issues was stimulated by exciting work being done on Aristotle's biology, which Allan has done so much to foster. I was privileged to attend a National Endowment for the Humanities seminar on Aristotle's biology, organized by Allan Gotthelf in Williamstown in 1983. It was a memorable and intellectually pivotal event for me, and I take this occasion to thank him. This essay is a descendant of Gill 2003, but substantially revised and expanded, including the discussion of Z.12. The current version profited from discussion at the Universities of Rochester and Oxford. I especially thank Jim Lennox for his helpful criticisms of the penultimate draft.

[1] Most scholars identify only two problems of unity in H.6 – forms and material composites – and take one or the other to be the chapter's main focus. I regard the unity of categorial properties (H.6.1045a36–b7) as a further topic.

[2] For Aristotle man is a universal compound composed of form and matter taken universally, not a pure form (*Metaph.* Z.10.1035b27–30). We might therefore wonder why he focuses on man rather than a pure form such as soul. I take it that he focuses on man as form, both here and in the related discussion in Z.12, because he is contrasting his own view with Plato's, and for Plato man is a sample form. Cf. Harte 1996, 281–2.

definition?[3] The question threatens the Platonists, who regard animal and biped as separate forms. Aristotle poses the problem thus:

Why is man not those [animal and biped], and men will exist by participation, not in one thing, man, but in two things, animal and biped, and so in general, man would not be one but more than one, animal and biped? It is evident that those who proceed in this way, as they are accustomed to define and speak, cannot give an account and solve the difficulty. (1045a17–22)

The trouble with the Platonists is that they treat their forms as definite and actual entities;[4] so the form man is what it is by partaking of two more ultimate entities, biped and animal. Aristotle states his own solution as follows:[5]

But if, as we say, the one is matter, the other form, and the one in potentiality, the other in actuality, the thing we are seeking would no longer seem to be a difficulty. (1045a23–5)

Apparently, biped animal is one thing and not two, because one of the components is matter, the other form, the one in potentiality, the other in actuality. But what does this proposal mean? The obvious place to look for clarification is in the next lines, where Aristotle discusses the unity of a material composite and says that the difficulty is the same.[6]

I will argue that the subsequent passage is the wrong place to look for clarification. We will properly understand the unity of composites only once we understand the unity of genus and differentia, and for that we must rely on Aristotle's first discussion of the topic in *Metaphysics* Z.12. But

[3] Aristotle discusses various sorts of priorities. In *Metaph.* Z.1.1028a34–6, he claims that Y is logically prior (i.e. prior in *logos*) to X, if the account of X must mention Y.

[4] Sometimes Aristotle complains that the Platonists' forms are *this*es (*tode ti*) (Z.14.1039a30–3). I am not here concerned with the question of whether his treatment of the Platonists is fair.

[5] Most scholars take the passage, as I do, to state Aristotle's proposal about the unity of form. Halper 1989, 182, who thinks that H.6 focuses exclusively on the unity of material composites, takes it to be part of that solution. See Harte 1996, 281 for objections to Halper's interpretation. On Harte's own view (291) this statement is the framework for Aristotle's solution to the unity of form – which she takes to be the main issue of the chapter – and the solution comes, not in the discussion of the unity of the composite (as is usually thought), but in a later passage on the unity of categorial items (1045a36–b7). My main objection to this proposal is that the later passage proves too much. If Aristotle is solving the problem of the unity of form, it is odd that he spends most of the paragraph talking about non-substantial properties – qualities, quantities, and so on. His focus suggests that the passage serves a purpose other than the one Harte attributes to it.

[6] Ross 1924, II.238 thinks that the analysis of the bronze sphere is analogous to that of genus and differentia. Rorty 1973, 411 thinks the two problems are literally the same: the genus animal is the stuff out of which a member of that genus is made. Ultimately I will argue that the two cases are analogous, but unlike Ross and others, I take the direction of the analogy and clarification to go from the unity of form to the unity of the composite, not vice versa.

before we turn to that, let us consider what he goes on to say, and why it fails to throw light on his proposal about the unity of form.

THE BRONZE SPHERE

Immediately after proposing his solution to the unity of form Aristotle says in H.6:

For (*gar*) this difficulty is the same, even if the definition of cloak is the "spherical bronze" (*ho strongulos chalkos*). For let this name ["cloak"] signal the account, so that the question is "What is the cause of the sphere and the bronze being one thing?" There no longer appears to be a difficulty, because the one is matter, the other form. So what is the cause of this, of the thing in potentiality being in actuality (*energeiai einai*), except the maker (*to poiēsan*), in those things for which there is generation? For there is no other cause (*aition*) of the sphere in potentiality being a sphere in actuality (*energeia einai sphairan*), but this (*tout'*) was the essence (*to ti ēn einai*) of each of the two. (1045a25–33)

This passage argues that a bronze sphere is one thing, because its components are related to each other as matter (the bronze) to form (the spherical shape). Notice that Aristotle does not explicitly say, as he did in the previous passage, that the matter is the thing in potentiality, the form the thing in actuality. Instead he asks "What is the cause of the thing in potentiality being in actuality?" The thing in potentiality is the matter, but the thing in actuality appears to be the composite – what the matter becomes when acted upon by an appropriate agent. Although Aristotle does not explicitly say so in this passage, the form can also be specified as the thing in actuality. At the end of the chapter he gives the following summary:[7]

But, as we have said, the proximate matter and the form are the same and one, the one in potentiality (*dunamei*), the other in actuality (*energeiai*). (1045b17–19)

The form and the composite are both properly designated as something in actuality: the form is the actual feature the matter potentially has, and the composite is the actual object the matter potentially is – the object the matter will become if it acquires the formal feature. Aristotle asks: "What is the *cause* of the thing in potentiality (the matter) being in actuality (the composite)?"[8] His answer is the form, but his statement is puzzling.

The final sentence announces that there is no other cause of the potential sphere's being an actual sphere, but this was the essence of each of the two.

[7] Cf. H.2, which focuses on form as actuality (*energeia*).
[8] For the question and Aristotle's reply, cf. Z.17.1041a21–b9.

What is the referent of "this"? What essence is he talking about? And what are the two things the referent of "this" is the essence of? Given the grammar of the sentence, the antecedent of "this" should be the cause (*aition*), which Aristotle also calls the maker (*to poiēsan*).[9] Scholars resist this identification, because they assume that the maker is the agent (e.g. a sculptor), and the agent is the wrong sort of thing to be an essence. But the maker need not be an agent. In *Metaphysics* Z.7 Aristotle says: "Indeed the maker (*to poioun*) and that from which the motion of becoming healthy begins, if the process is from art, is the form in the soul" (1032b21–3), and the same chapter identifies the form in the soul of the agent as the essence of the product (1032a33–b2).[10] We can therefore take the maker to be the form in the soul of the agent – the spherical shape, the blueprint or plan, which guides the sculptor's production.[11] If the sculptor's production succeeds, that form is later realized in the finished product, the bronze sphere.[12] We can now understand the spherical shape as the essence of both the composite bronze sphere and its matter, the bronze: the shape is the essence of the finished bronze sphere, since that shape makes the object what it is, a sphere. The shape is also the essence of the bronze, insofar as the bronze potentially has a spherical shape. The spherical shape is the goal of the potentiality and determines what the potentiality is a potentiality for. If this interpretation is correct, the definition of a bronze sphere mentions two things, bronze and sphere, but the two are intimately related, since the spherical shape is the essence of both the potential and actual sphere. The sphere once generated is one thing, a bronze sphere, and its definition is one, because the object defined is one. Thus the problem of unity dissolves.

But the problem does not in fact dissolve. The bronze has various potentialities grounded in its own nature as bronze (bronze is a hard, malleable stuff, which can be made into a sword, a plowshare, a triangle, and so forth, as well as a sphere). The bronze is a sphere in potentiality, but it is also actually bronze, and its essence as bronze is distinct from the spherical shape it potentially has. So the definition of the sphere as "the spherical bronze" mentions two distinct things after all, which are logically prior to the concrete whole. The notions of potentiality and actuality have simply been mapped onto a more basic hylomorphic framework, and they

[9] Cf. Charles 1994, 87–93 and 87 n. 21, though he would disagree with my identification of the cause and the maker.

[10] Cf. Z.7.1032b11–14; Λ.4.1070b22–35; *Ph.* 3.2.202a9–12.

[11] I have made this suggestion before, Gill 1989, 169–70. Aristotle sometimes calls the form/art responsible for change a first (unmoved) mover, something that moves other things, without itself being moved: *GC* 1.7.324a24–b14.

[12] This idea is well discussed by Modrak 2007, 113–14.

mask the problem.[13] If we ignore the supplemental machinery, Aristotle's proposal matches his treatment of matter and form in *Metaphysics* Z: the bronze which composes a sphere is a definite subject (*hupokeimenon*), and the spherical shape, which makes the bronze into a sphere, is a property (*pathos*) of it.[14] Since the shape can be stripped away, leaving the bronze intact, and can be realized in other materials, the shape is an *accidental* property of the bronze: the shape belongs to the bronze, but the bronze and the shape are defined independently of each other.[15] So the account of a bronze sphere mentions two components, sphere and bronze, the first predicated of the second, as *this in that* (*tod' en tōide*), form in matter.[16] In Platonic language, the matter *partakes* of the form.[17]

Some scholars think that Aristotle's real interest in H.6 is the unity of form and the *functional* matter of a living organism – a body organized in such a way as to support the psychic functions.[18] Although living organisms are doubtless Aristotle's central concern, he has a good reason to focus on a bronze sphere in H.6.[19] Organic compounds are subject to the same problem as bronze spheres, though less obviously. The situation is more complicated, because Aristotle thinks that organic matter lasts only as long as the composite it constitutes: if the form is removed, the matter is destroyed together with the composite. As he puts it, organic matter separated from the whole exists in name only:[20] a severed hand is not a real hand, because it can no longer operate as a hand does. Because the organic

[13] For the mapping, see *Metaph.* H.2.1043a14–21. This point is made by Loux 1995b, 251.

[14] This conception of the relation between matter and form is rooted in Aristotle's treatment of change in *Ph.* 1.7, where he argues that all changes involve a continuant, a subject that persists while losing one property and acquiring another. Aristotle relies on that model in his discussion of generation in Z.7–9 and H.1–5.

[15] In *APo.* 1.4, to be discussed below, Y is an accident of X, just in case Y is predicated of X, but Y need not be mentioned in the account of what X is, nor need X be mentioned in the account of what Y is.

[16] See, e.g., Z.8.1033a24–34, 1033b12–16, 1034a5–8; Z.11.1037a29–33; H.2.1043a4–7. See also passages cited in nn. 22 and 23 below.

[17] Loux 1995b, 271 attributes to Aristotle an attractive Fregean conception of predication according to which the subject and predicate fit together without a predicative link. So Loux would disagree with me that Aristotelian predication can be reformulated in terms of participation. But in fact Aristotle sometimes uses Platonic terminology, especially when contrasting some other relation to accidental predication, e.g. at Z.4.1030a11–14 and Z.12.1037b13–21 (I will discuss both passages below). Usually Aristotle speaks not of the relation of subject to predicate, but of that of predicate to subject: the predicate is *said of* (*legetai*) or *predicated of* (*katēgoreitai*) or *belongs to* (*huparchei*) the subject. He often uses prepositions: the predicate is *of* the subject (*kata* + genitive) or *in* (*en* + dative) the subject.

[18] Halper 1989, 188–94; Kosman 1984, 136; Lewis 1995, 244–9.

[19] In his response to Halper 1989 and Kosman 1984, Loux (1995b, 253–6, 261–5) emphasizes Aristotle's use of this example.

[20] See, e.g., *Metaph.* Z.10.1035b24–5; *GA* 1.19.726b22–4; *Mete.* 4.12.389b31–390a2, 390a10–13.

body depends for what it is on the soul of the organism, the human soul is an *essential* property of the human body, and the body ceases to be a human body when the soul is removed.[21] Despite the difference between functional matter and ordinary stuff such as bronze, functional matter is itself an accidental compound:[22] the functional properties belong to some lower-level matter, which persists when the animal dies. If the account of the composite must mention the form and the persisting matter, the composite is defined with reference to two more basic entities.[23] Aristotle says different things in different places about the precise level of the continuant. Non-uniform parts such as hands and feet are realized in uniform stuffs such as flesh and bone, and these are ultimately made of the four elements, earth, water, air, and fire (*PA* 2.1.646a12–24). Z.10 twice says that a human being perishes into flesh and bones (1035a17–19; 1035a31–3), suggesting that these uniform parts can outlast the organism. *Generation of Animals* 2.1, on the other hand, includes flesh with parts that are destroyed when the organism dies (*GA* 2.1.734b24–31). Flesh is the organ of touch, and it loses that capacity when removed from a living system. These complications do not affect the main point: at *some* level of hylomorphic analysis – whether at the level of flesh and bone, or at the level of the four elements – there is a subject to which the organic form belongs as an accident, and that subject survives its removal.[24] So the problem of unity is relocated at a lower level. Aristotle focuses on a bronze statue in H.6 to highlight the fact that both the problem and its solution concern the *continuant*, the persisting matter, whatever its level. Organisms are no better off than bronze spheres, when we consider the matter that constitutes them and survives their destruction.

The analysis of a bronze sphere resembles that of a white man, Aristotle's favorite example of an accidental compound.[25] In Z.4 he uses the same label "cloak" for a white man and asks whether cloak has an essence and

[21] As we shall see below, Y is an essential property of X, just in case Y is predicated of X and must be mentioned in the account of what X is. For the classic discussion of the different relation between form and matter in organisms and artifacts, see Ackrill, 1972/3.

[22] See Z.11.1036b21–32, where Aristotle criticizes a proposal by Socrates the Younger. In *Metaphysics* Z he defines composites as *this in that*, form in matter, even when the matter has a functional characterization.

[23] Aristotle claims that the composite bronze sphere is defined with reference to two things, form and matter, at *Metaph.* Z.7.1033a1–5; he extends the claim to living organisms and their kinds at Z.8.1033b24–6; and he denies that such kinds – universal composites – are substances at Z.10.1035b27–30.

[24] On this issue, see Lewis 1994, 273. For a critique of Lewis's position, see Rhenius 2005, 83–111.

[25] See Bostock 1994, 282. As Loux 1995b, 250 rightly points out, the machinery of potentiality and actuality, which we are ignoring, can be superimposed here too: a man, whatever his color, is potentially white, and when he has that property, he is actually white. Furthermore, whiteness, the color, is actually white.

definition. He gives the compound a single name, "cloak," but argues that the object lacks an essence and that its account is not strictly a definition.[26] Only those entities are strictly definable, he claims, which are said "not by predicating one thing of another" (*mē tōi allo kat' allou legesthai*) (1029b22–1030a11). A bronze sphere, like a white man, violates this rule, because it consists of two entities, which are logically independent of each other, and one of which is predicated of the other – the spherical shape of the bronze. Thus a bronze sphere is defined with reference to two more basic entities.

Far from explicating the unity of genus and differentia analogy with a bronze sphere puts Aristotle squarely in the Platonic position he was trying to avoid, that genus and differentia are two distinct entities, one (biped) predicated of the other (animal), as the spherical shape is predicated of the bronze.[27] Calling biped animal by the single name "man" conceals the problem, just as does calling a bronze sphere or a white man "cloak." This comparison does not solve the problem of the unity of form.[28] We should look again to Z.4: Aristotle explicitly distinguishes species of a genus from accidental compounds, saying that only species of a genus have an essence. Species, unlike accidental compounds, do not involve the participation (*ou kata metochēn*) of an object in an accidental property (*pathos*) (1030a11–14). Since Z.4 drives a wedge between the species of a genus, which have an essence and definition, and accidental compounds such as a white man or a bronze sphere, which do not, the assumption that H.6 clarifies the unity of genus and differentia by appeal to such a compound is a mistake.

Aristotle investigates the unity of definition in *Metaphysics* Z.12, and that is the place to look for help with the unity of form in H.6. Here he explicitly distinguishes genus–differentia combinations from accidental compounds. After we have considered Z.12, we will bring our conclusions to H.6 and then reassess the passage about the bronze sphere and its relation to the proposed solution for biped animal.

26 There may be some extended sense of "essence" and "definition" which allows us to call the account of a white man a definition (Z.4.1030a17–b13).

27 See Bostock 1994, 282.

28 Loux 1995b interprets H.6.1045a25–33, as I just have (though we differ on the question of a predicative link), but because he ignores the first puzzle about the unity of biped animal he does not discuss how the puzzle about the bronze sphere relates to it or what Aristotle means by calling the puzzles "the same" (1045a25). Lewis 1995 recognizes the first puzzle but does not address the difficulty it poses for his view. See n. 46 below. Bostock 1994, 282–3 takes the problem as evidence that Aristotle abandons genus–differentia definitions. For objections to Bostock's proposal, see Harte 1996, 282–4.

THE UNITY OF FORM IN Z.12

Metaphysics Z.12 presents itself as Aristotle's first attempt to explain the unity of definition (Z.12.1038a34–5), and here he explicitly denies that a genus is related to its differentia in the way that a particular white man is related to whiteness. Again he uses the example of man defined as "biped animal":

I mean this difficulty. Why is the thing one whose account we call a definition – for instance, man whose account is "biped animal"? Let that be its account. Why is this [man] one and not many, animal and biped? For in the case of a man and white they are many when one item does not belong to the other, whereas they are one when it does belong, and the subject, the man, is somehow qualified (*pathēi ti hupokeimenon, ho anthrōpos*) (for then the white man comes to be and is one thing). But in the other case [biped animal] one item does not partake of the other: the genus does not seem to partake of its differentiae (because [in that case] the same thing would at the same time partake of opposites, since the differentiae which mark off the genus are opposites). (1037b10–21)

If the genus partook of its differentiae, it would have incompatible properties – for instance, plane figure would be both round and square.[29] So the relation between a genus and differentia is not that between a subject and a property it has. Aristotle contrasts genus–differentia combinations with a white man, but he could equally have used the example in H.6, a bronze sphere. Both a white man and a bronze sphere feature two distinct items, a subject and a property, which constitute one object when the subject partakes of or has the property. The two components are logically independent of each other, and the property is an accident of the subject.

What, then, is the relation between a genus and a differentia, if the genus does not partake of the differentia? Aristotle discusses definition by division and then mentions two possibilities:

We must first consider definitions by division. For nothing else is in the definition except the so-called first genus and the differentiae. The other genera are the first and the differentiae taken together with it – for example, the first [genus] is animal, and the next is biped animal, and again wingless biped animal; and similarly even if it [the definition] is stated through more [terms]. And in general it makes no difference whether it is stated through many [terms] or few; so it also makes no difference whether it is stated through few or through two. Of the two, one is the differentia, and the other is the genus – for example of biped animal, animal is the genus and the other <biped> is the differentia. So if (1) the genus simply does not

[29] Cf. Z.14.1039b2–6, for a similar objection to the Platonic account.

exist apart from the species (*eidē*) of a genus, or (2) if it exists but exists as matter (for sound is a genus and matter, and the differentiae make the species (*eidē*) and elements (*stoicheia*) from this), it is evident that the definition is an account from the differentiae. (1037b27–1038a9)

Consider the two alternatives. The first recalls the *Categories*: the genus is simply more general than its species, so mentioning it adds no information that is not already contained in the species (*Cat.* 5.2b7–14). If one asks of a primary substance – say, a particular man – what it is, it is more informative to mention the species man than the genus animal, since man contains more information. *Metaphysics* Δ.28 offers a similar conception: plane is the genus of plane figures (square, circle, etc.), and solid is the genus of solid figures (cube, sphere, etc.). Aristotle says that the genus is the subject (*to hupokeimenon*) for its differentiae (*Metaph.* Δ.28.1024a36–b4), but presumably – if the account in Δ matches Z.12 – the subject-genus does not partake of them. Think of the genus as an indefinite determinable, which the differentia determines into a determinate species. On this conception the genus does not exist at all apart from the species of a genus, since its content is simply more general than theirs.

Aristotle's second alternative compares the genus to matter. Since this option is distinct from the other, the genus is apparently something over and above its species. He gives a helpful example: sound is genus and matter, and the differentiae produce the species and elements from this. The example recalls Plato's *Philebus*, where sound is used to clarify a certain type of division. The analytical method proposed in the *Philebus* is not the familiar method of division used in Plato's *Sophist* and *Statesman*. Those dialogues practice dichotomous division, the division of a kind into two sub-kinds (for instance, knowledge into practical and theoretical) in order to define some definite kind, such as statecraft, at the bottom of a genus–species tree. The *Philebus*, by contrast, analyzes a kind into its parts and features in order to clarify the kind itself, which is divided.[30] The investigator aims to figure out precisely how many (*hoposa*) units the kind is divided into and of what sorts (*hopoia*) they are (*Phlb.* 16c5–17e6).

Plato's Socrates gives two examples, both concerned with sound. Sound which comes through the mouth is one thing, but it is also unlimited in plurality (*Phlb.* 17b3–4). Our job is to determine how many and of what sorts the sounds are (*Phlb.* 17b6–9). A page later he returns to the same example, voiced sound, and this time treats it initially not as a single entity, as he did at the start of the previous passage, but as an unlimited

[30] I discuss the *Philebus* passage in Gill 2009.

plurality – a stream of undifferentiated vocal sound, which someone might experience. Socrates says that the Egyptian god Theuth discovered the letters by recognizing in the stream of vocal sound a spectrum voiced–unvoiced. He grouped together similar sounds into smaller and smaller ranges, until he singled out the indivisible letters, some voiced (the vowels), some intermediates (semi-vowels and others), and some unvoiced stops (the consonants) (*Phlb.* 18b6–d2). Thus he worked out how many indivisible sounds there are and of what sorts they are. In both versions of the example sound is grasped by one of its features, voice.

Socrates gives a second example, again to do with sound, but this time from a different perspective. Sound can be grasped by its pitch, and in this case the continuum is high, medium, and low tone. That continuum can be divided into smaller and smaller continua and finally into individual musical notes and intervals (*Phlb.* 17b11–e6).

Plato's examples reveal that sound can be considered from various perspectives. In focusing on one feature – pitch – we ignore other features of the kind, such as voice, volume, and rhythm.[31] Each feature is a range which admits a more and a less, and indivisible units are located somewhere within that range. The continuum is never used up: it can be extended indefinitely in either direction and can be divided into units in more than one way. The twenty-four letters of the Greek alphabet do not exhaust the spectrum of voiced sound,[32] nor do the notes and intervals in Greek music exhaust the continuum of pitch. Furthermore, the original choice of differentia (say pitch) ignores other features of the genus (voice, volume, rhythm). Thus the genus conceived as matter contains information not captured by any one division. So on the second conception the genus is something more than any one species of the genus.

In Z.12 Aristotle insists that proper procedure takes a differentia of a differentia, say biped as a differentia of footed, perhaps cloven biped as a differentia of biped, and so on down to indivisible forms of footed. He calls the final differentia the substance (*ousia*) of the thing (1038a19–20, 25–6). Once the differentiation is complete, all intermediate differentiae can be ignored: mention of footed is redundant once the lowest differentia is singled out.[33] Mention of the genus, on the other hand, is not redundant, since the content of the genus is not used up in any single line of division.

[31] Plato calls attention to these other features at *Phlb.* 17d4–6 and 25c9.

[32] The international phonetic alphabet includes more than a hundred possible sounds, of which ancient Greek recognizes only some.

[33] Cf. *Phlb.* 16c10–e2.

How, then, do we make sense of the unity of genus and differentia on the second conception?

Consider a division of sound into the letter alpha. The first genus is sound, which is immediately conceived under one aspect, voice, a spectrum marked off as voiced–unvoiced. That spectrum can be subdivided into smaller continua, voiced–intermediate–unvoiced. The letter alpha is located somewhere on the voiced end of the spectrum, distinguished from all other voiced sounds, perhaps by the openness of the mouth in voicing it.[34]

In the terminology of *Metaphysics* H.6, a genus such as sound is potential because it can be determined in various ways, in terms of voice, pitch, volume, and so forth. The feature selected at the outset, say voice, limits sound to one parameter. Although other features of sound, such as pitch, speed, and volume, characterize instances of a lowest kind, those other features stand outside the division of voice and are accidental features of individual voiced sounds. Sound is a nice example, because the indivisible kinds – for instance, the letters – can be adequately defined by a single line of division. Pitch, volume, and speed are not essential features of the phoneme alpha, the indivisible kind, though they characterize individual expressions of it.

Examples that particularly interest Aristotle – biological kinds such as man – are much more difficult, since they are determined as what they are by many aspects of the genus, not just one.[35] Scholars often point out that Aristotle's account in *Metaphysics* Z.12 and H.6 is overly schematic, suggesting as it does that the species of a kind can be fully determined by a single line of division. *Parts of Animals* 1.3 announces that correct procedure requires division straightaway by many differentiae (*PA* 1.3.643b9–26) – not merely by mode of locomotion, but by other functions as well, such as mode of feeding, reproduction, perception, and cooling.

Experts on Aristotle's biology now generally agree that his project was not to classify animal kinds into sub-kinds and species but to collect and explain animal differentiae, grouped under the general headings "ways of life (*bioi*), activities (*praxeis*), character-traits (*ēthē*), and parts (*moria*)" (*HA* 1.1.487a11–12; 491a8–11).[36] Given his procedure, a single type of animal will turn up in many distinct divisions, because it shares features with animals in different groups. Differentiae shared by groups of animals are explained

[34] See Smyth 1984, §7.

[35] On definition in Aristotle's biology, see Balme 1987; Lennox 1987b, and 2010. On intricacies of the problem, see Gotthelf 1985b, 1987, and 1997a.

[36] Pellegrin 1982; Balme 1987; Lennox 1987a.

(e.g. blooded animals have lungs because they are land-dwellers), and anomalies are singled out for special attention (e.g. whales and dolphins have lungs but are water-dwellers).[37] What is striking about Aristotle's procedure from the perspective of Z.12 is that different lines of division apparently yield separate building-blocks.[38] These elements are correlated in groups of animals, and the task is to explain why they occur together.[39] At the stage of mere division, before explanation is undertaken, features of the same creature which are singled out in different divisions are logically independent of one another. This is so, even if a necessary connection between those features is later established. If this observation is correct, it explains why different axes of division can be investigated separately, and why building-blocks can be combined in such a variety of ways in different sorts of creatures. At the same time it explains why mention of the genus adds no information beyond that contained in the final differentia – the genus adds no information, because it is immediately considered from the perspective relevant to that division, as in the *Philebus* example of sound and voice: all other generic information stands outside that division.

Differentiae can be investigated independently, but eventually they need to be integrated. In *PA* 1.5.645b14–20 Aristotle says that each of the parts of the body exists for the sake of some action (*praxis tis*), and that the whole body is constituted for the sake of some complex action (*praxeōs tinos heneka polumerous*).[40] A saw is constituted for the sake of sawing; similarly a body is constituted in a way (*pōs*) for the sake of its soul (*PA* 1.5.645b17–19). This passage suggests that an organism's various functions and parts are organized and coordinated, some subordinate to others, for the sake of the

[37] *PA* 3.6.668b33–669a1, 669a2–13. On this topic, see Lennox 1987a, 110–11.

[38] Cf. Pellegrin 1985, 106, who claims that parts of animals serve as a sort of alphabet: they can be combined in numerous different ways in different animals, and the combination promotes survival in different environments.

[39] A central passage for Pellegrin 1985, 98, 103–4 is *Pol.* 4.4.1290b25–38, which compares governments to animals. Both sorts of entities have many parts, and there are as many species of each as the combinations (*suzeuxeis*) of the necessary parts. Pellegrin claims that, whereas the *Politics* holds out the possibility of defining biological species, the biological works never actualize the possibility. In his view Aristotle's zoology is primarily moriology, the study of parts. Lloyd 1990, in a critique of Pellegrin (9–15), marshals evidence, especially from the *Metaphysics*, to show that Aristotle regards whole animals as substances and downgrades their parts, e.g. at *Metaph.* Z.16.1040b5–16 and 1041a3–5. Lloyd's criticism is fair, but Pellegrin has uncovered an important facet of Aristotle's biological project. Prior to investigating the unity of a species the biologist must locate the building-blocks, different combinations of which occur in different sorts of animals. To discover the unity of a species, one must explain why certain building-blocks are present and not others. Whether or not Aristotle carried out the further step of establishing the unity of species, he was evidently serious about the possibility, as witnessed by the passage in *Pol.* 4.4.

[40] Or *plērous*: complete. On this passage see Lennox 2001a, 176–8.

animal's way of life (*PA* 1.5.645b28–646a1).[41] The organization is in a way for the sake of the soul, because the soul controls the animal's characteristic behavior. Discovering the unity of a species is a teleological project: the features, first studied apart from one another, are together *hypothetically necessary* for the life of the organism they serve – they are jointly necessary *if* there is to be a creature that lives a certain sort of life.

Division, which precedes such an investigation, has a more modest aim, to identify functions and associated parts shared by groups of animals. Those animals may be quite diverse in other respects. This project can be carried out piecemeal without knowing how or why functions fit together as they do in particular sorts of animals. Z.12 gives a schematic account relevant only to division. The chief point is that on either conception of the genus a mention of it adds no information that is not contained in the final differentia. On the first conception the genus is nothing over and above its species; on the second conception, though the genus is something more, it is immediately conceived from a point of view relevant to the division undertaken, and that is the only aspect of the genus relevant to that division.

Aristotle does not decide between the two conceptions in Z.12, but in *Metaphysics* H.6 he claims that the puzzle about the unity of genus and differentia is solved by regarding the genus as matter and the differentia as form. Thus he takes up the second interpretation – genus as matter, like sound.[42] He now adds that the one is in potentiality, the other in actuality. The new machinery is not merely superimposed onto the old hylomorphic framework, as we originally thought. On either conception of the genus, the genus is potential because it is a determinable kind. On the second conception, the genus is immediately considered under one of its aspects, say mode of locomotion, and other determinations stand outside that line of division. Since only that aspect of the genus is considered which is relevant to the selected axis, the genus and differentia are one, even though the genus contains information not actualized by that division.

[41] On this topic, see Lennox 2010. See also Charles 2000, esp. ch. 12; Halper 1989, 114–18; and Modrak 2001, 189–94. The account is likely to be highly complex, as Gotthelf (1997b) shows in his analysis of the elephant's trunk.

[42] My current interpretation revises my previous view of the unity of form in H.6 (in Gill 1989, 138–44, and 168, and Gill 2008), in which I took Aristotle to invoke the first notion of the genus in Z.12. Given that view, I found a significant disanalogy between his treatment of genus and differentia and his treatment of matter and form in the next part of H.6. On my current understanding, the analogy is much closer.

THE UNITY OF THE COMPOSITE IN H.6

We observed that the passage on the bronze sphere, though often taken to clarify the unity of form, presents the wrong conception for that purpose. As Aristotle has construed the relationship between matter and form until now, the matter partakes of the form. That is the model he rejects for genus and differentia in Z.12. Given what he says about participation in Z.12, it is noteworthy that later in H.6 he criticizes those who appeal to participation to explain the unity of concrete composites:

> Because of this difficulty [about the being and unity of material composites such as bronze spheres and of non-substantial properties such as whiteness], some people speak of participation and are puzzled as to what the cause of participation is and what participation is; and others speak of communion, as Lycophron says that knowledge is [the communion] of knowing and soul; others speak of living as a composition or binding together of soul with body. In fact, the same argument applies to all – for indeed being healthy will be either a communion or binding together or composition of soul and health, and bronze being a triangle will be a composition of bronze and triangle, and being white will be a composition of surface and whiteness. The reason is that they seek a unifying account of potentiality and actuality (*entelecheias*) and a difference [between them]. (1045b7–17)

As I interpreted the example of the bronze sphere before, the analysis invites Aristotle's own criticism.[43] Indeed, he calls attention to the problem by mentioning a bronze triangle as one example of the wrong analysis (1045b14–15). On the view to be discarded, being a bronze triangle is a composition of bronze and triangle, two distinct things bound together by a unifying relation. On this conception, as we saw, the notions of potentiality and actuality are simply mapped on to the more fundamental matter–form framework without helping to solve the problem. True, a lump of bronze is potentially spherical, but it is also something actual in its own right. The actual identity of bronze makes it a hard, malleable

[43] I concur with Bostock 1994, 286, who takes the passage to discuss ordinary predication. The various designations – "participation," "composition," etc. – are different names for one basic relation. There has been an extraordinary effort on the part of commentators to interpret the passage in such a way that Aristotle's criticism does not apply to his own usual treatment of matter–form composites. I agree with Witt 1989, ch. 4 (and for a distinctive view along similar lines: Scaltsas 1994), that Aristotle's explication of matter and form in terms of potentiality and actuality in H.6 and Θ cannot be reduced to predication. But unlike others I take Aristotle's potentiality–actuality model in H.6 and Θ to replace the predication model used to explicate composites in Z and H.1–5.

stuff suited to take on the shape of a sphere: its actual identity grounds its potentiality.[44]

What, then, is going on in the passage about the bronze sphere in H.6? Notice that Aristotle speaks of the "spherical bronze" (*ho strongulos chalkos*) – an unusual phrase, which specifies the form with an adjective, and the matter with a noun[45] – rather than his more usual "brazen sphere" (*hē chalkē sphaira*), which specifies the composite with a noun and the matter with an adjective. "Spherical bronze" could recall the designation of an accidental compound, like "white man," but more likely it highlights the similarity with "biped animal" just discussed. I have argued that the passage about the bronze sphere does not spell out Aristotle's proposal about the unity of genus and differentia. He has already clarified the unity of genus and differentia in his second proposal about the genus in Z.12, which he recalled in H.6, elaborating it by appeal to potentiality and actuality. On my view he *extends* that proposal to the case of concrete material composites and will extend it again to the unity of categorial properties. We are to understand the unity of the composite on analogy with the unity of form.

An objection to my proposal is Aristotle's use of the particle γάρ ("for") at H.6.1045a25.[46] He turns from the puzzle about the unity of form to the puzzle about the unity of the composite, saying: ἔστι γὰρ αὕτη ἡ ἀπορία ἡ αὐτὴ κἂν εἰ κτλ. ("*For* this difficulty is the same even if the definition of cloak should be 'the spherical bronze' . . . " [1045a25–6]). Given the γάρ, readers expect the upcoming discussion of the composite to clarify the preceding discussion of form. But as we saw, if that is the point of the discussion of the bronze sphere, his solution is no more successful than that of the Platonists, whose view he rejects. Since he denies that the genus partakes of its differentiae (Z.12.1037b18–19), and since H.6 itself faults explanations in terms of participation, we should look for some other way to interpret the passage.

[44] Cf. H.4.1044a27–9, where Aristotle tells us that a saw cannot be made out wool or wood – the matter must be of a suitable kind to be made into the product. See also *de An.* 2.5.417a26–8: a subject has a first-level potentiality to have some property, if it is of a suitable kind to have it.

[45] Cf. Z.8.1033a32–3.

[46] The objection has been put to me by Lewis 1995, 258 n. 39, and Harte 1996, 282 n. 15. Lewis himself faces a problem in light of the *gar*. He says that the problem of the unity of form has a counterpart at the level of the compound material substance (234–5) and takes the *gar* to indicate that the solution for the composite explains the unity of form (258 n. 39). As he understands composites in his paper, at some level form is an accident of matter. Similarly, then, the differentia will be an accident of the genus. But that is precisely the idea Aristotle rejects in Z.12 when he denies that the genus partakes of its differentiae (1037b18–21). The unity of the genus and its differentiae is not clarified by comparing it to an accidental compound.

For a start, we need not read γάρ at 1045a25:[47] the sequence γ-α-ρ in the manuscripts can be reconstrued as γ' ἄρ, formed from γε ἄρα, with two elisions (the second before the initial vowel in αὕτη). This revision calls for no emendation of the transmitted text, but simply a different word division.[48] Denniston in his *Greek Particles* lists several examples of γ' ἄρ and γ' ἄρα in Plato, one in Democritus, as well as examples from poetry.[49] If the letters are construed here as I suggest, Aristotle is making an inference from his previous point, not clarifying or explaining it. He uses the unusual phrase, "spherical bronze," to spotlight the analogy with "biped animal": "spherical" specifies a differentia, and "bronze" a genus.

Consider again the passage I quoted earlier, this time reading γ' ἄρ:

> This difficulty is then (γ' ἄρ) the same, even if the definition of cloak should be the "spherical bronze." For let this name ["cloak"] signal the account, so that the question is: "What is the cause of the sphere and the bronze being one thing?" There no longer appears to be a difficulty, because the one is matter, the other form. So what is the cause of this, of the thing in potentiality being in actuality, except the maker, in those things for which there is generation? For there is no other cause of the sphere in potentiality being a sphere in actuality, but this was the essence of each of the two. (1045a25–33)

This passage claims that the problem of unity for concrete composites disappears. As I previously interpreted the passage, the problem did not disappear, because the continuant, bronze, was not only potentially a sphere, but also actually bronze, a definite subject. So the bronze sphere was defined

[47] In fact, even if read *gar*, the word can be construed as a confirmatory adverb, "in fact," "indeed," rather than as a causal conjunction. See Smyth 1984, §2803. I thank Trent Dougherty for reminding me of this possibility. Cf. Gill, 1995, 516 n. 7.

[48] Word division and diacritical marks are the work of Hellenistic editors; they were not used by the Classical authors.

[49] Denniston 1966, 43. It is generally agreed that γάρ was derived from γε and ἄρ (Denniston 1966, 56; cf. LSJ s.v. γάρ). There is often disagreement among the manuscripts between γάρ and γ' ἄρ', γ' ἄρα, or simple γε or ἄρα. Consider some examples of γε ἄρα: *Phlb.* 46a12: Σύμμεικτον τοῦτό γ' ἄρ', ὦ Σώκρατες, ἔοικε γίγνεσθαί τι κακόν ("That really seems to be a mixed experience *then*, with a bad component, Socrates"); *Tht.* 171c10–d1: Ἀλλά τοι, ὦ φίλε, ἄδηλον εἰ καὶ παραθέομεν τὸ ὀρθόν. εἰκός γε ἄρα ἐκεῖνον πρεσβύτερον ὄντα σοφώτερον ἡμῶν εἶναι ("But it is not at all clear, my dear Theodorus, that we are running off the track. *Hence* it is likely that Protagoras, being older than we are, really is wiser as well"). Cf. *Phlb.* 35b6. For discrepancy in the MSS, cf. *Chrm.* 159d4. Diels in earlier editions of *Fragmente der Vorsokratiker* printed the last sentence of Democritus, Fr. 191 with γ' ἄρ: ταύτης γ' ἄρ' ἐχόμενος τῆς γνώμης εὐθυμότερόν τε διάξεις καὶ κτλ. ("When you keep to this mind, *then*, you will both carry on in better spirits and . . . ") but the present edition (1951–2) prints γάρ. To the passages in Denniston should be added *Tht.* 152c2–3: οἷα γ' ἄρ' αἰσθάνεται ἕκαστος, τοιαῦτα ἑκάστῳ καὶ κινδυνεύει εἶναι ("*So* it looks like things are for the individual such as he perceives them"). McDowell 1973, 110 adopts this reading from Badham in his translation and commentary, as does Levett in Burnyeat 1990, 272. See also Lee 2005, 80 n. 7. The new Oxford Classical Text of Plato, vol. 1 (Duke *et al.* 1995) retains γάρ.

with reference to two more basic components, the bronze and the spherical shape.

We can keep the main contours of the previous interpretation: the form (spherical shape) explains why a sphere in potentiality is a sphere in actuality, and it is the essence of both the potential and actual sphere. But I modify the previous interpretation in one fundamental respect. What makes the problem of unity disappear is the treatment of matter and form on analogy with the treatment of genus and differentia in the preceding passage. That difficulty vanished, because the genus is a determinable kind, like sound, which the differentia determines into a determinate kind, like the letter alpha. The genus does not partake of its differentiae but is actualized in a particular way by a single line of division. Although the genus contains information not captured by a single dimension, other features of the genus stand outside the species viewed from that perspective (e.g. man considered in his locomotive dimension). The analogous solution for a material composite is to treat the *matter* of the spherical bronze in the way one treats the *genus* of man. The bronze that constitutes the sphere is then not a definite subject to which the shape belongs. Instead, the bronze is something determinable, which the spherical shape differentiates into a sphere, much as biped differentiates animal into man (*qua* locomotive). Call this determinable matter the *generic matter*. To understand the import of Aristotle's analogy, we need to look at his discussion of matter and potentiality in *Metaphysics* Θ.7.

But first we should consider Aristotle's further extension of the genus–differentia model to non-substantial properties in H.6.[50]

UNITY OF NON-SUBSTANTIAL PROPERTIES

Aristotle turns to categorial properties – substantial kinds, qualities, quantities, and other properties.[51] In contrast to material composites,[52] entities that have no matter are, he says, straightaway some one thing. Each categorial item is some one thing, just as it is some being, whether a substantial

[50] In the final lines of the section on composites Aristotle extends his point about matter and form to mathematical objects, such as circles, which contain intelligible matter (1045a33–5). His solution applies to anything that contains matter, whether perceptible or intelligible. Cf. Z.10.1036a9–12, on intelligible matter. Harte 1996, 287–9 argues convincingly against Ross 1924, II.238, that intelligible matter is the abstracted matter of mathematical objects, such as two-dimensional magnitude, not a genus.

[51] Like Halper 1989, 184–6, and Harte 1996, 289–90, I take Aristotle to be discussing the unity of items in the categories, not the unity of the categories themselves, contrary to Ross 1924, II.238.

[52] Including intelligible composites. See n. 50 above.

kind (*tode*),[53] a quality (*to poion*), or a quantity (*to poson*) (1045a36–b7). The claim about non-substantial properties is surprising for readers of *Metaphysics* Z. In Z.1 Aristotle declared that qualities such as white and other non-substantial properties are defined with reference to something in the category of substance.[54]

Aristotle's view in Z appears to rely on a distinction in *Posterior Analytics* 1.4 between two sorts of predicates that belong to a subject in virtue of itself (*kath' hauto*) (*APo.* 1.4.73a34–b5).[55] A predicate Y belongs to a subject X *kath' hauto* in one way, if Y is predicated of X, and Y must be mentioned in the account of what X is. For instance, animal belongs in this way to man, since animal is predicated of man and must be mentioned in the definition of man; and heat belongs in this way to fire, because heat is predicated of fire and must be mentioned in the definition of fire. Let us call Y an *essential* predicate of X. More precisely:

> Y is an *essential* predicate of X, if and only if Y is predicated of X, and Y must be mentioned in the account of what X is.

A predicate Y belongs to a subject X *kath' hauto* in a second way, if Y is predicated of X, and the subject X must be mentioned in the account of what Y is. For instance, female belongs to animal *kath' hauto* in the second way, because female is predicated of animal, and animal must be mentioned in the account of what female is. Similarly, odd belongs to number *kath' hauto* in the second way, because odd is predicated of number, and number must be mentioned in the account of what odd is. To invoke Aristotle's favorite example of a *kath' hauto* predicate of the second sort, snubness belongs *kath' hauto* to nose, because snubness is predicated of the nose, and the nose must be mentioned in the account of what snubness is. Snubness (a quality) is defined as "concavity in a nose," as "this in that" (Z.5.1030b14–20). Let us call Y a *special* predicate of X.[56] More precisely:

> Y is a *special* predicate of X, if and only if Y is predicated of X, and X must be mentioned in the account of what Y is.

[53] Substantial kinds, such as the species man and horse and the genus animal, were treated in *Metaphysics* Z as universal composites, which contain form and matter taken universally. See passages cited above in n. 23.

[54] Z.1.1028a20–9; Z.4 and 5 are largely devoted to non-substantial properties and their secondary status; see esp. Z.4.1030a17–b13; Z.5.1030b14–28 and 1031a1–14.

[55] For the distinction, see also *Metaph.* Δ.18.1022a14–19, 1022a24–32. The whole investigation of categorial being in Z–H is the study of kinds of *kath' hauta* being. See Δ.7.1017a22–4, to which Aristotle refers in the opening sentence of Z.1 (1028a10–13).

[56] I owe the label to Paul Coppock.

According to the *Analytics*, Y is an *accidental* predicate of X, just in case Y is predicated of X, but Y need not be mentioned in the account of what X is, nor need X be mentioned in the account of what Y is. Whiteness is a special predicate of surface, but an accidental predicate of a man.

According to *Metaphysics* Z every non-substance is a special predicate of some definite entity in the category of substance, which is its primary recipient, and with reference to which the non-substance is defined and understood. Thus health is a special predicate of living thing, and justice is a special predicate of man. In Z.5 Aristotle claims that snubness and things like it are defined "from addition," because their account must mention the sort of subject in which the property is always realized (Z.5.1030b23–8). Non-substances, though not themselves composites of matter and form or substance and property, are nonetheless defined with reference to something outside their own category in the category of substance. In Z.4 and 5 Aristotle argues that these entities lack the requisite unity of primary things and so are not strictly definable.

He still holds something like that view at the beginning *Metaphysics* Θ, where he sums up the discussion in Z–H before turning to the main topic of that book, potentiality and actuality:

> We have talked about *what is* in the primary sense and to which (*pros ho*) the other predicates of being are referred, namely, substance (for the other beings are stated according to [*kata*] the account of substance – quantity and quality and the other things said in this way – for all will contain the account of substance, as we said in our first remarks). (Θ.1.1045b27–32)

If H.6 is consistent with Θ.1, the claim about the unity of non-substances is not that they can be defined apart from substance. Female is still defined with reference to animal, and whiteness with reference to surface. The passage that follows in H.6, in which Aristotle criticizes people for their appeal to participation and the like (1045b7–17), bears directly on the discussion of non-substances. The passage mentions not only a bronze triangle, thus recalling the discussion of material composites. The main examples are manifestations of non-substances, such as being healthy and being white. On the model to be rejected, being healthy (*to hugiainein*) is a binding together of soul and health, and being white (*to leukon einai*) is a composition of surface and whiteness.[57]

[57] Bostock 1994, 286 complains that the definitions that Aristotle states look somewhat circular (e.g. being white is a composition of surface and whiteness). That problem is mitigated, if Aristotle is talking about the manifestation of non-substances – the equivalent in the non-substance categories of concrete composites in the first category.

Aristotle's proposal about the manifestation of categorial properties again extends his solution for the unity of form. Occurrences of whiteness and snubness, which are defined with reference to some subject in the category of substance, involve a subject conceived of as matter and a predicate conceived of as form, the one in potentiality, the other in actuality. The subject is a determinable kind, and the outcome of differentiation is the manifestation of some categorial item. Thus being white (an occurrence of whiteness) is surface (a determinable kind) differentiated by whiteness (a quality). Surface has many potentialities (to be hard or soft, rough or smooth, white or black), but all aspects of surface other than the potentiality to be white or black are external to the differentiation of surface into an instance of whiteness.[58] Any such instance will occur in a surface which is also smooth or rough, hard or soft, and so on, but those features merely coincide with the whiteness in the surface. Again, being female is defined with reference to animal, but the only aspect of animal internal to its differentiation as female is the reproductive capacity. All other features of animal – mode of feeding, locomotion, perception, and so forth – are external to being female, even though every instance of female occurs in an animal which has features of those sorts as well. So interpreted, the genus, though it contains information not actualized in a differentiation, does not undermine the unity of the item differentiated. The definition of a non-substance still mentions a substantial kind, but the reference does not add information, since the kind is immediately considered from the perspective of that feature.

MATTER AND POTENTIALITY

H.6 ends with a summary of Aristotle's proposals:

But, as we have said, the proximate (*eschatē*)[59] matter and the form are the same and one, the one in potentiality (*dunamei*), the other in actuality (*energeiai*), so

[58] Note that at Z.4.1029b21–2, Aristotle supposes that a white surface can be defined as a smooth surface. He is relying on Democritus' view, which he criticizes at *Sens.* 4.442b11–19.

[59] The Greek *eschatē* can be translated either "proximate" or "ultimate," depending on whether one is counting up from the bottom or down from the top. I have vacillated in my translation of it in this passage, rendering it with most translators as "proximate" in Gill 1989, 143, but as "ultimate" in Gill 2008. I became dissatisfied with "proximate," because the proximate (or highest-level) matter in organisms is the functional matter, which is not the issue in H.6. The relevant matter in H.6 is the matter which is potentially the product, in the sense that it can be turned into the product without further changes of it (Θ.7: see below). This is the proximate matter, as opposed to lower-level matter, which must first be transformed into suitable stuff. Although organisms have functional matter, they also have proximate matter, which separates out when the organism dies. That is the sort of matter Aristotle is talking about in H.6.

that it is like seeking what is the cause of oneness and of being one; for each thing is some one thing, and the thing in potentiality and the thing in actuality are somehow one, so that the cause is nothing else unless there is something that moves it from potentiality to actuality. And all those things that have no matter are simply just some one thing. (1045b17–23)

This passage indicates that Aristotle has in mind some difference between a pure form such as man, defined by genus and differentia, and a concrete material composite such as a bronze sphere, since he says that the latter is *somehow* one (*hen pōs estin*), whereas things without matter – evidently including non-substantial properties – are *simply* some one thing (*haplōs hoper hen ti*). What is the difference? To answer this question and to understand more fully Aristotle's proposal about composites, we need to look at his treatment of matter and potentiality in *Metaphysics* Θ.7.

Metaphysics Θ as a whole seeks to clarify the notions of potential and actual being, and Θ.7 focuses on matter and potentiality. The chapter breaks into two main parts.[60] The first part asks when an entity is properly called potential and argues that something is potentially in some end-state when it is sufficiently worked up that it can be in that state without further changes to it – nothing needs to be added to the subject, subtracted from it, or otherwise changed (1049a5–12). Earth is not yet potentially a sphere, because it must first be combined with water and worked up into bronze, but once the combination has been transformed into bronze, there is stuff of a suitable kind to be a sphere. The potentiality to be in the end-state is grounded in what the subject is, its own actual identity.

The second part of the chapter focuses on the product, and the question is this: "Does the entity designated as potential in the first part of the chapter persist in the product as something *actual* as well as *potential*?" Aristotle answers that it depends on what sort of continuant and what sort of change we are talking about. If we are talking about a man who comes to be healthy, musical, or white, then the persisting subject, the man, remains actually a man when his potentiality to be healthy or musical or white is realized. The complex, say a white man, is a subject – man – characterized by an accidental property (*tois pathesi to hupokeimenon anthrōpos*, 1049a29). The subject has some actual identity in its own right: it is a *tode ti*, a particular thing of a definite sort (1049a27–30).

[60] I discuss Θ.7 in more detail in Gill 2008.

But the situation is different in the case of matter and form. Aristotle characterizes the form–matter relation as follows:

> In cases not like that [i.e. not like a physical object and its non-substantial properties], but the thing predicated (*to katēgoroumenon*) is some form (*eidos ti*) and *this* (*tode ti*), the proximate [subject] (*to eschaton*) is matter (*hulē*) and material substance (*ousia hulikē*). And calling [a thing] "that-en" (*ekeininon*) with reference to its matter and its properties (*kata tēn hulēn kai ta pathē*) turns out to be quite correct, since both are indefinite (*aorista*). (1049a34–b2)

Form–matter predication is not ordinary predication. The item predicated is said to be some form and *this* (*tode ti*), and the matter of which it is predicated is variously described as *indefinite* (1049b2), *potential* (1049a18–24), and *not* a *this* (1049a24–9). We specify the product adjectivally (as "that-en") with reference to it.

The important idea is that the matter is present in the product only potentially and not actually. In *On the Heavens* 3.3 Aristotle defines a corporeal element as that into which other bodies are divided, which is present in the complex either potentially or actually, and he adds that it remains disputable in which of the two ways this occurs (*Cael.* 3.3.302a15–18). These two alternatives – matter as actual and matter as potential – reflect the distinctive treatments of matter in *Metaphysics* Z and H.1–5 and in H.6 and Θ. In *Generation and Corruption* 1.10, he pursues the second idea in his analysis of mixture and argues that the ingredients of a mixture exist actually before they are combined but are only potentially present in the compound (*GC* 1.10.327b19–31). Ingredients of the original sort can be extracted by destroying the mixture, and in that sense they are potentially present, but they are not actually there in the compound.[61] Aristotle was not an atomist. A physical analysis of compounds will not uncover particles of the stuffs that went into the mixture. Even so, the presence of those stuffs is felt in the mixture, because the compound has certain properties owing to its ingredients. Bronze, a compound of copper and tin, has the color, the strength, the malleability, and other dispositional properties it has because

[61] For a different interpretation of mixture, see Lewis 1994, 272–5. Loux (1995a, 500 and 1995b, 260–1) raised objections to my interpretation (in Gill 1989, 168–70) and I replied in Gill 1995, 517–19. I would now add to my earlier response that Loux's objection assumes that some *thing* ("it") persists throughout a substantial generation but goes shadowy while it composes a higher composite. That is not what I think. On my view, the original stuff is gone, transformed into the higher composite. When the higher composite is destroyed, matter of the original sort is left behind. This is not numerically the same stuff that was there at the outset, but merely stuff of the same sort (cf. *GC* 2.11.338b16–19: my use of the word "recreated" [1989, 170] was misleading). The higher composite is connected to the matter from which it was generated, because certain properties of that matter characterize it.

of the metals used in its composition. It shares some properties with its ingredients, but its own essential features, which differentiate it from them, are not the same as theirs. If constituent matter is not a definite and actual subject to which the form belongs, but something merely potential, this also justifies Aristotle's claim that it is not a *tode ti* – not a definite thing of a certain sort.

What is the nature of matter's indefiniteness? Aristotle claims that a thing is specified adjectivally with reference to both its non-substantial properties (*pathē*) and with reference to its matter. Thus we call a sphere "white," not "whiteness," and we call it "brazen," not "bronze." This usage is correct, he claims, because both are indefinite (1049a36–b2). Non-substances have categorial content and so are quite definite in that respect. Their indefiniteness is rooted instead in their dependence on some definite object for their instantiation. Whiteness when it occurs is a property of some definite thing, say a man. Matter as conceived of in Θ.7 resembles non-substantial properties in its indefiniteness. The bronze which can be made into a sphere has its own character as bronze, but once it is worked up into a sphere, the bronze depends for its existence on the object whose matter it is. The material properties of a sphere, which connect it with its simpler origins, are not properties of the constituent bronze as an independent actual subject, but properties of the sphere it constitutes. That is why Aristotle approves of Greek usage, which prefers adjectives to nouns in specifying an object with reference to its matter.[62]

Very important to understanding matter in Θ.7 is Aristotle's claim that an object is called "that-en" with reference to the matter at only the next level down, not with reference to anything lower (1049a18–22). For instance, a box is called "wooden," but not "earthen," even though wood is itself a compound of earth and water. This claim corresponds to the thesis in the first half of the chapter, that matter is potential only once it has been sufficiently worked up to be in the end-state without further changes of it. If matter were an ordinary subject, the essential properties of the lower-level matter should also apply to stuffs and objects further up the series. It should therefore be acceptable to call a box "earth-en," since

[62] It is instructive to compare Aristotle's discussion of the same linguistic practice at Z.7.1033a5–23. There he gave a not very convincing story about why people use adjectives instead of nouns to specify an object's matter. In fact ordinary usage conflicts with the account of matter he was exploring and apparently defending in Z. *Metaphysics* Z treated the relation between form and matter on the ordinary predication model. If bronze is a definite and actual subject of which a spherical shape is predicated, it should be quite correct to call the statue "bronze," just as we call a musical man "man." In Z.7 Aristotle was trying to rationalize a practice that was at odds with his theory. In Θ.7, by contrast, he commends the practice which properly reflects his theoretical proposal.

earth is an elemental constituent of a box. If Aristotle were an atomist, he should have no objection to calling a box "earth-en," or for that matter even "earth." Atoms combine to form molecules; molecules combine to form more complex structures. The atoms are still there in the complex structure. This is not Aristotle's view, as is evident from his denial that a box is specified adjectivally with reference to matter more remote than its own proximate matter. At each stage the lower-level matter, which is potentially something at the next level up, is transformed into the entity at the next higher level. So it is correct to specify the product adjectivally with reference to the matter at the first level down, but incorrect to specify it adjectivally with reference to any matter below that.

An advantage of Aristotle's proposal in Θ.7 is that a statue is not *two* things with different persistence conditions, the bronze and the statue, but just *one* thing: the statue. The generic matter of the statue is a collection of properties that belong to the statue. The material properties that connect an object with its simpler origins account for certain aspects of its behavior. For instance, a statue's constitution from bronze explains its heaviness, solidity, and meltability. But those material properties do not contribute to its essence: the essence of a composite is its form.[63]

At the end of H.6 Aristotle says that material composites are *somehow* one in contrast to forms and categorial properties, which are *simply* one. The difference is that material objects contain matter, whereas forms and categorial properties do not (1045b23). How do material properties weaken the unity of the object to which they belong?

I have argued that the proximate matter of an object resembles an ordinary genus on Aristotle's second conception in Z.12. An ordinary genus, too, has features not captured by a particular differentiation, and those features characterize the differentiated object, though they are accidental to it insofar as it is conceived from the perspective of that determination. A material genus is similar, since features of the matter are either hypothetically necessary (hence explained by the form) or accidental to the object as the thing that it is. Even so, an object's material properties differ in a fundamental respect from its other properties, because the material properties can essentially characterize a simpler stuff into which the higher object

[63] Jim Lennox reminds me that Aristotle often defines animals with reference to their matter as well as their form. But I believe that Aristotle's arguments in H.6 and Θ.7 enable him to define animals with reference to their form alone. Material properties of an object are either hypothetically necessary, and so follow from the form, or accidental. In either case they need not be mentioned in the animal's definition. Aristotle may of course sometimes define animals with reference to their material parts ("biped animal" is a good example), but such definitions can be reformulated in terms of psychic functions (e.g. locomotive capacity).

degenerates. For instance, coldness and dryness are essential properties of earth, and coldness and wetness are essential properties of water. So the material features tend to undermine the unity of the object to which they belong. The elements are worked up into complex material objects, but they tend to separate out of the whole and to behave in the manner of those simpler stuffs.[64]

Given that material properties undermine the unity of the object to which they belong, does Aristotle overstate his case in saying that composites are *one*, even if only in a way (*pōs*)? As I understand his view, material composites are genuine unities, but their unity is unstable, and remaining a unified whole demands considerable work. Material objects have characteristic activities (animals nourish themselves, grow, reproduce, perceive, move about in their environment as they do) and these activities are controlled by their form. An object's characteristic activities are not merely expressions of what it is; those activities also maintain and renew it, preserving it from degeneration, destruction, and loss of form. Any material object composed of Aristotle's four elements eventually wears out, degenerates, and is destroyed into simpler stuffs. That is why material composites, though genuine unities while they last, are only somehow one. Their unity is temporary and eventually gives way.

64 On the undermining role of matter, see *Metaph.* Θ.8.1050b6–28; *Cael.* 1.12.283b19–22, and *Cael.* 2.6.288b15–18.

CHAPTER 6

Definition in Aristotle's Posterior Analytics

Pierre Pellegrin

In 1981 the proceedings of a Symposium Aristotelicum on the *Posterior Analytics* held in Padua three years earlier were published. These papers, which were noteworthy, have had a great influence on the exegesis of the Aristotelian corpus up to the present day. Two of them contributed more than the others to the shaping of a new conception of the *Posterior Analytics*, which, in my view, is still alive and active, namely the papers by Jonathan Barnes and Jacques Brunschwig. It is very likely that when they were published, these papers were considered, by their authors as well as by their readers, to be remarkably divergent. But it is a quite frequent effect of temporal distance to reveal convergences which were not previously obvious. What are these convergences, and which new elements did they introduce?

Previously, the main problem interpreters were concerned with was to locate the *Posterior Analytics* within the chronological and/or logical framework of the Aristotelian corpus, and to find the relation it has to the other treatises of the *Organon*. One may recall, in this connection, the divergent views of Heinrich Maier (1896–1900) and Friedrich Solmsen (1929). Maier thought that Aristotle, having first held, in the *Topics*, a Platonic position, which was used as a basis for criticizing previous philosophers – 'sceptical' philosophers according to Maier (Sophists, Megarians, Anthistenes) – later departed from Platonism in the *Prior Analytics*, where he exposed his own method, namely syllogistic, and finally restricted himself to a special case of syllogism, demonstration, in the *Posterior Analytics*. According to Solmsen, on the contrary, Aristotle first developed his doctrine on demonstration in the *Posterior Analytics* before he built up a 'general syllogistic' in the *Prior Analytics*.

This chapter is dedicated to my dear friend Allan Gotthelf, who has accompanied me on the byways of Aristotelian biology.

The 1981 papers mentioned above also ask the kinds of question that were asked by Maier and Solmsen. But it is within the *Posterior Analytics* itself, rather than between the *Posterior Analytics* and the rest of the *Organon* that they purport to discover some differences, if not inconsistencies. It is worth mentioning that Barnes's and Brunschwig's papers were written against the background of questions about the internal chronology of Aristotle's treatises: 'Was syllogistic, which is the subject of the *Prior Analytics*, discovered before or after apodeictic, which is the subject of the *Posterior Analytics*?' 'Which of these treatises was written first?' – questions which bring these papers close to the works of Maier and Solmsen, but which seem today less attractive and which are considered by most scholars to be undecidable. But the means by which Barnes and Brunschwig reach their goal are quite interesting. Barnes shows, in a very convincing way, that one can to a large extent uncouple syllogistic and apodeictic. As Barnes puts it, 'although the *Analytics* presents Syllogistic and Apodeictic as a single system, the former a necessary propaedeutic to the latter, the Syllogism is in fact an incidental adjunct to the theory of demonstration'.[1] I agree with Barnes that the core of the epistemological doctrine of the *Posterior Analytics* does not require knowledge of the twists and turns of syllogistic, which are exposed in the *Prior Analytics*. Barnes goes even further in claiming that the Aristotelian apodeictic is, so to speak, viable without any reference to syllogistic. In this respect, Barnes considers two of the requirements for the premises that make, in Aristotle's view, a demonstration scientific immediacy and universality. These requirements seem to be taken from the developments in syllogistic in the *Prior Analytics*, but Barnes shows that even in the case of universality, apodeictic can fly with its own wings, without any help from syllogistic. As I have noted elsewhere,[2] in spite of some 'local' difficulties we may have in harmonizing his analyses with the text of the *Posterior Analytics*, Barnes is basically right. Understanding, and even elaborating, the theory of demonstration found in the *Posterior Analytics* does not require a great knowledge of syllogistic, and, what is more important, it does not require a theoretical commitment to syllogistic. This has an important historical consequence that Barnes strongly underlines. It is well known that, according to the opponents of Aristotelian science – and this is true from the Renaissance onwards – recourse to the syllogism had been an obstacle to the development of the sciences. If one adopts Barnes's position, Aristotle's theory of demonstration does not inherit the scientific vices of syllogistic. The most interesting example of such a freeing of Aristotelian

[1] Barnes 1981, 33. [2] Pellegrin 2005, 17.

epistemology from a tutelage by syllogistic is that of mathematics. One may be surprised that Aristotle did not see that syllogistic form was not adapted to the exposition of mathematical reasoning. If the Aristotelian theory of demonstration is independent of the logic found in the *Prior Analytics*, it would not leave aside the only really formalized science in ancient Greece, namely geometry.

The paper by Brunschwig shares with that by Barnes the same chronological perspective, but Brunschwig insists much more on the gaps that are internal to the *Posterior Analytics*. In doing so, he actually provides Barnes's position with a much firmer basis. Brunschwig writes,

> Il est donc permis de penser que les deux objets qu'Aristote assigne à ses *Analytiques*, au début comme à la fin de son ouvrage, sont bien distincts, et que deux entreprises discernables coexistent ou se superposent dans les *Seconds Analytiques*. L'une, par rapport à laquelle l'objet d'étude se détermine sous le nom de science démonstrative, vise essentiellement à situer celle-ci dans le cadre d'une théorie générale de la science, et à décrire ses liens d'opposition et de complémentarité avec la science des principes. L'autre, par rapport à laquelle l'objet de l'étude se détermine sous le nom de démonstration, vise essentiellement à situer celle-ci dans le cadre d'une syllogistique générale.[3]

Brunschwig concludes that the second 'enterprise' presupposes a general theory of the syllogism, that is the *Prior Analytics*, whereas the first one does not. And – must we say 'unfortunately'? – Brunschwig draws from this some chronological conclusions: what Aristotle says on demonstrative science could have been written *before* the other 'enterprise', which is about demonstration, especially as, Brunschwig adds, 'its problematic has clearly enough its origin in the central books of the Platonic *Republic*'.

I would like to test those readings, and especially Brunschwig's, on a particular point, namely definition. For it seems to me that two different series of texts in the *Posterior Analytics* give two images of definition which, if not inconsistent with each other, at least are different on some crucial points. We therefore have to go through these texts, tackling some well-known textual difficulties. Finally, my chapter is more about the *Posterior Analytics* than about theory of definition, properly speaking.

[3] Brunschwig 1981, 79 ('It is, then, allowed that the two aims that Aristotle assigns to his *Analytics* at the beginning, as at the end, of his work, are quite distinct, and that two discernible enterprises coexist or are superimposed in the *Posterior Analytics*. The one, in relation to which the object of investigation goes by the name of "demonstrative science", looks to situate itself within the framework of a general theory of science and to describe its connections of opposition and of complementarity with the science of principles. The other, in relation to which the object of investigation goes by the name of "demonstration", looks essentially to situate itself within the framework of general syllogistic').

DEFINITION AS A PRINCIPLE

In *Posterior Analytics* 1.2 and 1.10, Aristotle considers the different types of principle on which scientific syllogisms are founded. Chapter 2 is entirely devoted to a kind of catalogue of the properties that make a syllogism a scientific demonstration. If this is to be the case, the cause of the demonstrable thing must be known, it must be the case that this cause be the cause of the thing, and this thing must be necessary. Then Aristotle considers the first premises a science is based on, that is to say the principles, since 'I say that first premises and principles are one and the same thing' (72a6). These principles must have some properties, be 'true, primitive, immediate, better known than the conclusion, prior to and the cause of this conclusion' (71b21).

At 72a7–24 Aristotle gives a list of the different types of principle. This is a difficult and controversial text, which has often been corrected and that we, consequently, have to look at with some care. Aristotle immediately affirms that principles are *propositions*, a claim which has to be reconciled with the passages in which he seems to say that principles are entities. In fact, this problem, which has been often raised by scholars, may be, if not solved, at least skirted. That is what I have done in my translation, understanding *archē d' estin apodeixeōs protasis amesos* (72a7) not as a definition of the term 'principle' ('a principle of a demonstration is an immediate proposition', Barnes 1994), but as a description of principles which happen to be propositions: when, in a demonstration, a premise is immediate, it is a principle.[4] The following section of the text gives a series of definitions related to each other. A proposition is a kind of a statement[5] which attributes one predicate to one subject. A statement is one part of a contradiction, a contradiction being 'a pair of opposites between which, in their own right, there is nothing' (72a12). In other words, there can be affirmative as well negative propositions. Then Aristotle divides principles

[4] I translate: 'est principe une prémisse immédiate d'une démonstration'. It should be noted that, in spite of its translation first by 'premise' and then by 'proposition', the term *protasis* does not really change its meaning between 72a7 and 72a8.

[5] Literally 'a part (*morion*) of a statement'. As Barnes notices (1994, 98), *pace* Mignucci (1975, 33), a proposition cannot be part of an *apophansis* which is itself 'a part of a contradiction' (72a11). Barnes, following Colli 1955, proposes to read *antiphaseōs*, 'contradiction' for *apophanseōs*, 'statement' at 72a8. This would be perfect, if Aristotle's text was not a series of definitions. Now Aristotle first defines *apophansis* and later *antiphasis*. Therefore the original text at 72a8 does have *apophanseōs*. The best solution is perhaps to take the term *morion* at 72a9 in the general sense of 'subdivision' (cf. *Metaph.* 5.25.1023b18: 'we say that species are parts of a genus'); hence my translation by 'sorte': a statement may be of two sorts, affirmative and negative. Philoponus seems to have had a reading of this kind (cf. 32.15; Wallies 1909). When, in either of these two cases, a predicate is attributed to a subject, we have a proposition. In any statement, one of the two sorts, either affirmative or negative, is true.

into two classes, axioms and posits, the latter being divided into hypotheses and definitions. In my view, this double division must be understood in the following way. An axiom is necessary to learn 'anything' (72a16), whereas a posit is an indemonstrable proposition that it is not necessary to grasp to learn 'something', that is, anything. The posit has to be grasped only when we want to learn something belonging to the same genus as the posit. In fact we find here a distinction, often mentioned by Aristotle, between common and proper principles.[6]

Concerning the two kinds of posit, the text is far from clear. Let us have a look at it:

> A posit which assumes either of the parts of a contradiction – I mean, for example, that something is or that something is not – is a hypothesis; without this it is a definition. (72a18–21)

Most commentators take it that there are two kinds of posit, hypotheses, which are existential propositions ('X exists', 'Y does not exist'), and definitions, which are not ('X is a biped animal'). Against this reading, it may in the first place be objected that the Greek text does not favour it as clearly as it at first seems. For, even if the expression τὸ εἶναί τι ἢ τὸ μὴ εἶναί τι (72a20) may mean 'that something is or something is not', it does not necessarily refer to existence but may also mean 'the fact of being something or not being something', a reading which seems to have been that of manuscript C, which omits the second τό – see Zabarella (672 F): 'esse vel non esse aliquid', and Mignucci (1975, 36–7). This sentence may also refer to an attribution within a predication, as Philoponus says (35.4: 'a hypothesis affirms or denies something of something'). But it is mainly the phrase 'without this' (*aneu toutou*, 72a21) which has to be considered carefully, and on which we must probably depart from the common interpretation. It seems to me that this does not refer to 'that something is or that something is not', as commentators usually suppose, this expression being just an example (cf. 'I mean for example', 72a20), but has to be referred to 'which assumes either of the parts of a contradiction'. Therefore it is not a relation to existence which differentiates hypotheses from definitions, but the fact that a definition cannot be negative.[7] Thus, this passage, in my view, says that definitions cannot be negative propositions. This affirmative

[6] The distinction in *APo* 1.2 leaves aside the case of axioms common to several sciences, but not to all of them.

[7] Mignucci 1975, 37, though he does not take τὸ εἶναί τι ἢ τὸ μὴ εἶναί τι in an existential sense, nevertheless takes 'without this' to refer to this expression. He therefore understands that Aristotle says that definitions may have a non-predicative form. But Aristotle explicitly affirmed that he was dealing with principles that are propositions (72a7–8)

character of definitions fits well with the quite central role they have among proper scientific principles. If, actually, scientific knowledge does share the characters mentioned above – necessity, indication of the cause, premises with specific properties – if so, how could there be negative hypotheses which are principles?[8]

Let us take a further step. In chapter 1.6, Aristotle intends to show that demonstration remains within the sphere of the '*per se*', an expression that was defined in chapter 1.4. Demonstration demonstrates *per se* and not accidental attributions. But what interests us here is the kind of relationship the premises have to a '*per se*' attribution. It seems to me that it is not the same '*per se*' attribution that we find in the premises and in the conclusion of a demonstration.

Chapter 1.6 begins with an argument, the invalidity of which has been recognized by scholars for a long time. If science, Aristotle says, starts from necessary principles, and if the *per se* attributes are necessary, then a demonstrative syllogism starts from premises in which the predicate belongs *per se* to the subject.[9] This fallacy is, all in all, not very important for what is at stake:[10] in the premises of a demonstration, the subject and the predicate have a *per se* attribution. The relationship between the subject and the predicate in the conclusion is also a *per se* attribution. This is stated in this same chapter 1.6: 'of accidents that are not *per se* in the way in which "*per se*" was defined, there is no demonstrative science' (75a18). 'The way in which "*per se*" was defined' obviously refers to chapter 4, in which Aristotle exposes the different meanings of the phrase '*per se*'. But there is a difficulty which has not been noted by many interpreters. Chapter 1.4, to sum it up in few words, distinguishes four senses of the phrase '*per se*'. The first two involve the essence of the subject (the line belongs *per se* to the triangle because it belongs to its essence; 'odd' belongs *per se* to number because the latter belongs to the essence of the former), the last two, on the other hand, refer to the opposition between substance and accidents for the first one, and, perhaps, to a causal relationship for the other (substance is '*per se*' because it is not said of a subject; when an animal died being

8 In a 'dialectical' passage, as the entire chapter 2.3 is intended to be, Aristotle considers it as obvious that definitions are affirmative: 'definitions seem to be of what something is, and what something is is in every case universal and affirmative' (90b3).

9 The text literally says: 'from some things of this kind' (74b10), which Ross seems to interpret as meaning 'from necessary premises', which would make the argument circular, its conclusion being identical with its major.

10 Perhaps the minor premise of the argument is implicitly considered as convertible. In this case, we would have: 'All necessary attributes belong *per se* to their subject', and we should reduce the sense of 'necessary' to the cases in which it means '*per se*'; but, if this is so, we should wonder what this argument is aimed at. Perhaps there is no syllogism in proper form in this passage.

sacrificed, the relationship between the sacrifice and the death is '*per se*'). That, in the case of the conclusion of a scientific syllogism, the phrase '*per se*' has to be understood in one of the first two senses has been taken by scholars to be so obvious, that, according to them, the only relevant question is whether the first or the second is concerned. So R. McKirahan,[11] who proposes arguments, which are quite impressive indeed, in favour of the first sense. Now, as Barnes, who is one of the few commentators aware of this difficulty, remarks, the *per se* accidents, namely the attributes in the conclusions of demonstrations, cannot fall under any of the first two senses of the phrase '*per se*', because they involve the essence of the thing, and not its accidents, even its *per se* accidents. In the proposition which constitutes the conclusion of the scientific syllogism, it is therefore impossible either that the subject includes the predicate in its own definition (first sense of '*per se*'), or that the subject belong to the essence of the predicate (second sense of '*per se*'). Only the premises of the scientific syllogism are concerned with the first two senses of '*per se*'. We shall have to deal again later with this difference, which is constitutive of the Aristotelian science, between the premises of the scientific syllogism, which remain within the sphere of essence, and the conclusion, where what is at stake is not the essence, but the properties of the essence.

The only attributions that can be *per se* in the first two senses of chapter 1.4 are the attributions of a predicate to a subject which is not the result of any demonstration. It is therefore not sufficient for a proposition to be a premise in a scientific syllogism, if this proposition is to display a *per se* attribution in one of the first two senses; it must also be the case that this premise be not the result of a previous demonstration. For, if this premise displays the *property* of an essence as has been deduced through a previous syllogism, the subject and the predicate of this premise cannot have a *per se* relationship in any of the first two senses of chapter 1.4, because these senses do not apply to properties. Only the principles of sciences may have such a *per se* attribution. In this respect, definitions, as they are principles, might, because of their very status, include a *per se* attribution in one of the first two senses. In my view, we have here a crucial point: in a scientific syllogism which is based on premises that are at the same time principles, the premises and the conclusion do not display the same kind of *per se* attribution.

As a principle, a definition is mainly characterized by its *position* in the deductive schema of science. Before any demonstration, which is a

[11] 1992, 169–71. Cf. also Ross 1949, 530. Mignucci does not mention this problem.

demonstration of a property, an essence must be posited of which the demonstrated property will be a property. In this respect, a definition happens to be known without being deduced from something else; through a demonstration, on the contrary, new things, namely properties, come to be known. But Aristotle has explored another approach to the relationship between definition and demonstration, which is not the cooperative relation just mentioned, but a relation of concurrence. This new approach, which focuses on the similarities and differences between definition and demonstration, is concerned with the pretension of definition to be a mode of knowledge in addition to demonstration.

DEFINITION AS A METHOD FOR KNOWLEDGE

This alternative approach to definition is already present in book 1 of the *Posterior Analytics*,[12] but it is developed in book 2, from chapter 3 to chapter 10. When it is considered as a mode of knowledge, definition is the knowledge of the 'what it is'. Now the 'what it is' is one of the four objects of inquiry listed in 2.1. The problem is, then, to know whether, given that demonstration has been recognized as the adequate means for scientific knowledge, the 'what it is' is knowable by way of demonstration or not. The question of the borderline between demonstration and the other modes of knowledge is all the more important in the case of the 'what it is', given that in chapters 2.1 and 2.2, Aristotle has shown two things. First, each of the objects of his inquiry has something corresponding to it in science. This is, at least, what is explicitly said in the first sentence of book 2: 'the things we seek are equal in number to those we know scientifically' (89b23). Now the 'what it is' is among the things we seek. Therefore there must be something which corresponds to it in scientific knowledge. Second, in 2.2 Aristotle establishes that the inquiry about the four objects he has listed in 2.1 can be reduced to the search for a middle term. Such a reduction induces us to consider that it might be possible to annex the inquiry about the 'what it is', which has to be a search for a middle term, to scientific demonstration.

Rather than the role of definition in science and its position in the deductive schema of science, chapters 2.3–10 consider the scientific capacities and pretensions of definition. Then the main question which interests Aristotle is to know whether we may get scientific knowledge of the 'what it is', a question that may be reformulated in this way: is it possible to demonstrate

[12] Cf. 1.8.75b31–2.

the 'what it is'? There is a preliminary question, on which Aristotle does not spend a long time, but which is of great historical importance, which is to know whether 'it is possible to know the same thing in the same respect both by a definition and by a demonstration' (2.3.90b2). Aristotle's answer is clear: if by 'to know' we mean 'to have scientific knowledge', then there is but one way to know something, which is to have a demonstration, namely a scientific syllogism of it. From a historical point of view, chapters 2.3–10 must be considered as one of the most vigorous and precise criticisms against the Platonic method of division, which was considered a method of definition. According to the members of the Academy, knowledge *par excellence* – and in this respect they were the true heirs of Socrates – was the construction of definitions. For Aristotle, on the contrary, to know in a scientific way is to demonstrate.

Let us, then, follow those chapters 2.3–10, gathering some remarks which may help us to characterize more accurately the new aspect of definition at this point of the *Posterior Analytics*. First of all we have to keep in mind that there is a radical break in the argument, explicitly indicated by Aristotle himself, between chapters 3 to 7, on the one hand, which consist in a 'diaporematic' inquiry, that commentators used to call a 'dialectical' one, and chapters 8 to 10, on the other, which adopt a more genuinely Aristotelian point of view. This means, among other things, that the first part may have recourse to arguments and may offer answers to some questions that are not totally endorsed by Aristotle.

There are several points that may interest us in chapter 3. Aristotle establishes that 'there are not definitions of everything of which there are demonstrations' and that 'there is no demonstration of everything of which there is a definition' (90b18). He intends first to show that the set of demonstrable things is not identical with the set of definable things. Later he will show that nothing is at the same time demonstrable and definable. Concerning the first question, the passage 90b24ff. is particularly informative. As it is sufficient, in order to topple the thesis Aristotle is criticizing, to find *one* case in which this thesis is false, he takes the case of the definitions which are principles. Principles, of course, are indemonstrable, therefore there are things which are definable but indemonstrable. There are definitions that are principles,[13] but Aristotle is very careful not to say that all are.

[13] Aristotle's formulation, 'the principles of demonstrations are definitions' (90b24) shows that what he has in view are the proper and not the common principles.

Chapter 4 draws the caricature of a demonstration which would demonstrate the 'what it is', specifying that the premises of such a syllogism are essential attributions and that they are also convertible. Then the syllogism does not produce anything which had not been previously posited. The true demonstration, on the contrary, leaves the circle of the essence without demonstrating this essence: 'it is not the case that if A follows from B and B follows from C, then A will be what it is to be C' (91b1). These remarks agree with what has been said on the difference, in book 1, between the *per se* predications in the premises which are principles and that in the conclusion. In the syllogism considered here, all the attributions, those in the premises as well as those in the conclusion, have recourse to one of the first two senses of '*per se*', that is, the senses which involve essence. I will not decide which of these two senses is here relevant. In no way can such a syllogism be a scientific one.

Chapter 5 has often been misunderstood, even by people who have a clear idea of the role of division in the grasping of definitions. Aristotle does not say whether he approves or disapproves of the use of division to find definitions. When, at 91b35, he writes that 'if you state a definition on the basis of a division, you do not make a syllogism', he says nothing on the validity of such a procedure. He just observes that some people grasp definitions in this way, and he affirms that in doing so they do not demonstrate. We know that these people were members of the Academy. There is but one object in this chapter: to show that division, as it is practised, does not demonstrate definitions. Much has been said on a possible connection between Platonic division and the Aristotelian syllogism.[14] This chapter is an essential element in such a debate, and it underlines the differences between division and syllogism in a more vigorous way than chapter 1.31 of the *Prior Analytics*, which considers division as a 'weak syllogism' (46a33). If, in some cases, division happens to establish a definition, this definition will in no case be demonstrated.[15]

Chapter 7, the last of the 'dialectical' part of the inquiry into definition in book 2 of the *Posterior Analytics*, includes some really difficult passages,[16] but its general scope is clear enough. As chapter 3 already did at 90b14, it introduces induction in the confrontation between definition and demonstration. From this, we can see, as Zabarella says, that,

[14] In a famous article, Paul Shorey intended to show that it is rather a passage in the *Phaedo* that led Aristotle to the discovery of the syllogism. See Shorey 1924.

[15] In the expression 'the syllogism proving by way of division' (6.92a28), the term 'syllogism' is taken in a broad and non-technical sense.

[16] Its last sentence, for instance.

contrary to demonstration and induction, definition is not 'an instrument for knowledge' (1098 E), or rather it is not a means for learning, as it is true that 'we learn either by induction or by demonstration' (1.18.81a39).

In chapter 8 Aristotle begins the, properly speaking, 'scientific' inquiry into the relationship between demonstration and definition. The enormous literature that has been produced on chapters 8 to 10 frees me from considering many interesting points in these chapters, to concentrate on what is relevant to my topic.

Aristotle begins by delimiting his inquiry, such a framing being a crucial point for my own analysis. Probably referring to 2.2.91a31, he first reminds us that 'it is one and the same thing to know what something is and to know the cause of whether it is' (93a4). We may note here that a part of the manuscript tradition has 'the cause of what it is', but this *lectio facilior* has to be rejected, because what interests Aristotle here is the cause of the thing and not the explanation of the essence, a question that will appear later, when several types of definition will have been distinguished. This is, in a way, what follows what is said in this passage: 'the reason of this is that there is a cause'. Barnes, in his commentary to the *Posterior Analytics*, says that this makes no sense ('I can make no sense of this', Barnes 1994, 217). He therefore proposes to put a comma after the *ti* at 93a5 and to understand 'the account of the fact that something is is the cause', which would be both a commonplace and repetitive of what has just been said. In fact Aristotle's intention is to restrict his investigation to the cases in which a scientific syllogism is possible, and, for this purpose, there must be a cause and such a cause must be the middle term of this syllogism. This is what is explicitly said at 93a7.

Then Aristotle clarifies the limits of his inquiry. We have at this point a problem of interpretation of the kind that commentators are especially fond of, because the cause is minuscule and the effect quite huge. Let us consider O. Goldin's translation of 93a5–8:

> The account of this is that there is some cause, and it is either the same or different, and if it is different, either demonstrable or indemonstrable. If now it is different and it is possible to demonstrate, it is necessary that the cause be a middle term and shown in the first figure. (1996, 102)

Relying on a strictly grammatical reading, commentators – and this, at least, from Philoponus onwards – understand, as Goldin[17] does, that after Aristotle has said that the presence of a cause has the result that it is one and

[17] Who, on the core of the interpretation of chapters 2.8–10, radically differs from Philoponus.

the same thing to know the essence of a thing and the cause of the existence of this thing, he next proposes two disjunctions: (i) the cause is either the same or different; (ii) this cause is either demonstrable or indemonstrable. I do not understand this passage in this way.

In the expression καὶ τοῦτο ἢ τὸ αὐτὸ ἢ ἄλλο (93a5), it is clear that τοῦτο refers to 'cause', but Aristotle does not want to say that the cause is either the same or different. As Ross puts it (1949, 628), together with 'old' interpreters who are accustomed to Aristotle's style, Aristotle makes here a distinction between a cause which is the same with the thing it is the cause of, and a cause different from the thing it is the cause of. 'Hæc causa vel eadem cum re, vel alia', Zabarella says (1108 A). Or, in other words, Aristotle makes a distinction between two situations: in one case the cause is the thing itself, in the other case the cause is something different.[18] Most of the commentators who adopt this interpretation – a correct one, I think – see at this point an allusion to the difference between substance and accidents, but Barnes is probably right that the allusion is to the difference between principles and propositions deduced from principles. For a principle, which is not deduced from previous propositions – because it is 'first' – cannot have a cause different from itself.[19] Now, as the last lines of the chapter will remind us (93b18–20), Aristotle wants to restrict himself to the case of the 'what it is' which has a cause different from itself, in other words to the 'what it is' the account of which is not a principle. But the account of the 'what it is' is precisely a definition. Therefore the definitions our passage is concerned with are no longer the definitions which are principles. We are, then, no longer looking at definitions from the perspective of book 1, in which Aristotle almost exclusively referred to definitions as principles. The short chapter 9 is entirely devoted to this very point: the definitions of the things which do not have causes other than themselves are principles that we *posit*, whereas our inquiry into the relationship between definition and demonstration is concerned only with definitions of things which have a cause different from themselves.

Commentators affirm that the second disjunction is between demonstrable and indemonstrable causes, and this leads them to a lot of difficulties, but also to exploits of imagination. Here again we have a remarkable example of Aristotle's syntactical off-handedness. In the expression κἂν ᾖ ἄλλο,

[18] J. Tricot, who is a perfect example of a translator accustomed to Aristotle's style, understands that 'the cause is either identical with the essence, or different from it', and this is what Aristotle has in mind. But Tricot (1962) understands the second disjunction (see below) as opposing 'demonstrable essence' and 'indemonstrable essence'.

[19] Cf. *APr.* 2.16.64b35: 'principles are known by themselves, and what comes after the principles through something else'.

ἢ ἀποδεικτὸν ἢ ἀναπόδεικτον (93a6), it is clear that if Aristotle had been doing a grammatical composition, 'demonstrable or indemonstrable' would have had to be referred to 'cause'. In fact, as Goldin remarks, it is also grammatically orthodox to refer these two adjectives to 'what it is', as Tricot does. But in a chapter that intends to decide whether the 'what it is' is demonstrable or not, it would be curious to begin by saying that the 'what it is' is 'demonstrable or indemonstrable'. If we construe the sentence according to its meaning, this second disjunction must be related to the *thing* about which we seek whether its 'what it is' can be demonstrated or not. Among the things that are not *causa sui* (first disjunction), some are objects of science (demonstrable), some are not (indemonstrable), for instance the contingent facts.[20] It is only in the case of demonstrations that it may be asked whether the 'what it is' of the subject of which *per se* attributes are demonstrated can itself be demonstrated. Aristotle, then, considers definitions that are not principles, and from which, if taken as premises, some properties can be demonstrated.

We have, therefore, to keep in mind that the definition which chapter 8 and almost the whole of chapter 10 deal with is not the definition which is a scientific principle, which is the subject matter of book 1 and of few passages in book 2. In the famous examples of eclipse and thunder, the definitions are not first principles. But, on the other hand, Aristotle also acknowledges, even in those chapters 3 to 10 of book 2, that there are definitions which are principles, and in chapter 9 he gives as an example of a principle the definition of unity, which is a principle of arithmetic (93b24). But in this case the 'what it is' is immediate (93b22).

DIFFERENT KINDS OF DEFINITIONS IN 2.10

After many others, I have now to consider the problem of the plurality of definitions in 2.10. This chapter is famous for the divergent interpretations to which it has given rise, the most obvious of its difficulties being that, in the course of the chapter, Aristotle seems to distinguish four kinds of

[20] I do not understand Ross's position: he reads this distinction in the way the other interpreters do and, one page below, takes a position very close to the one I propose here (1949, 629). For a similar construction, cf. 1.8.75b34, a passage about the phenomena that occur regularly, such as the eclipses: 'it is clear that in so far as they [the demonstrations] concern things of this kind, they exist for ever, but in so far as their objects do not exist always, they are particular'. The text literally says, in the last section of the sentence: 'in so far as they do not exist always, they are particular'; grammatically the subject should be 'the demonstrations'. Cf. Verdenius's commentary (1981, 347), in which he quotes Bonitz (*Index Aristotelicus* 239a24): 'ellipsis sujecti, quod, licet diversum sit a proximo, tamen facile suppleatur'.

definition, and then, in his final summary, he explicitly says that they are only three ('third', 94a13). A traditional solution, which can be traced back at least to Themistius (51.3–23),[21] is to hold that the first kind of definition – usually, and unfortunately, called 'nominal definition' – is omitted by Aristotle in his final summary. Why is this so? Because the nominal definition would not be a 'real' definition of something – an object or a natural class – really existing. Let us follow the text.

Chapter 10 begins in this way:

Since a definition is said to be an account of what something is, it is clear that one type will be an account of what its name, or some other name-like account,[22] means, for example what does [the answer to the question] "what is a triangle as a triangle mean?"[23] (93b29–32)

It is generally taken for granted that, according to Aristotle, one cannot know the essence, and therefore the definition, of a thing, if this thing does not exist: 'it is impossible to know what something is when we do not know whether it is' (8.93a20). Aristotle goes even further, since he requires, as a preliminary requirement for knowing the 'what it is', that we know 'something of the thing' *not accidentally*, such a remark being difficult to interpret with exactitude.[24] Commentators, from Aspasius to Ross, think that I know an eclipse accidentally if I know it through an accident belonging to some other genus, and this is certainly right. If, for instance, I know that when there is an eclipse dogs bark, this does not direct me towards the essence of the eclipse. But do I, at least, know that there is an eclipse when dogs bark? If yes, there would in effect be two necessary stages for the grasp of the 'what it is': first, to know that the thing exists (this would be possible even accidentally: there is an eclipse since the dogs bark), and, second, to possess about the thing non-accidental information (in the case of the eclipse that it is 'a sort of privation of light', 93a23). But this is not what Aristotle says. On this point,

[21] Wallies 1900.

[22] Ross 1949, 635 thinks that Aristotle has here in mind expressions such as 'straight line'; but Philoponus (362.25; Wallies 1909) considered this phrase as an equivalent to the expression at 7.92b28 'an account which means the same as the name'. This seems more likely to fit, for instance 'figure with three angles' for triangle (Philoponus 372.18; Zabarella 1132 E).

[23] Keeping the τί ἐστι of the manuscripts at 93b31. The manuscripts are further divided over two readings: τρίγωνον and ᾗ τρίγωνον. With Ross's bracketting at 93b31: οἷον τὸ τί σημαίνει [τί ἐστι] ᾗ τρίγωνον, the text means 'which "triangle" signifies', and this is certainly what Aristotle means. 'Triangle' refers to the term, otherwise it would not be an example of what came before.

[24] This requirement is really crucial and has been the subject of a huge literature; cf. in recent times Goldin 1996, 113ff. I will come back to this below, without interpreting completely the expression 'knowing something of the thing'.

chapter 8 is in a way contradictory, at least verbally. At 93a21 Aristotle writes: 'the "if it is," sometimes we grasp this accidentally, and sometimes by grasping something of the thing itself, for example . . . of an eclipse, that it is a sort of privation of light'. Against such a perspective, as I have just noted, the accidental grasp is actually a grasp: when dogs bark (at least in some cases), I know that there is an eclipse. But just after this passage, Aristotle says: 'when we know accidentally that these things are, necessarily we have no grasp of what they are, for we do not even know if they are' (93a24–6). When we have an accidental grasp of something, it is useless to try to follow the path towards the definition of this thing, because we do not even grasp the existence of this thing. Aristotle seems here to correct himself and to reinforce the requirements for the grasping of the existence of a thing. The knowledge that something is, which is prior to the knowledge of what it is, must be the knowledge of something proper to the thing. The barking of dogs does not help us to know anything of the eclipse, not even that there is an eclipse.

When we do not know the 'that it is', or the 'if it is', of something, we cannot grasp its 'what it is', and it makes no sense to seek its definition. This is what Aristotle already said in chapter 7, with an important additional point and a famous example:

> It is necessary that anyone who knows what a man or something else is also know that it is (for of that which is not, no one knows what it is. You may know what the account or the name signifies when I say 'goat-stag', but it is impossible to know what a goat-stag is). (92b4–8)

A very unfortunate parallelism in the mode of expression here had as an effect that interpreters have considered that the beginning of chapter 10 alludes to an example similar to that of the goat-stag. This is why many of them thought that what they (but not Aristotle) called the 'nominal definition' was a definition of the same kind as the 'definition' of goat-stag. But if we look carefully at the text, we see that this is not what Aristotle actually says, and that the parallelism between chapters 7 and 10, which is mainly based on the verb 'to signify', is an artificial one. In chapter 10 Aristotle deals with the definition which is 'an account of the "what it is"', and, within this genus, he distinguishes a first kind, the nominal definition. In chapter 7, on the contrary, he contrasts 'what the name signifies' with 'definition', that is, he does not include the signification of the name ('goat-stag') within the class of definitions: 'of that which is not no one knows what it is', that is, no one knows its 'what it is'. In no way is there a definition of any kind for the goat-stag.

We may use other words to say the same things. Chapter 10 says that, among definitions, some just signify, whereas some others 'prove'.[25] It is worth remembering the context within which one can find the goat-stag example. It appears in a chapter in which Aristotle develops the idea that demonstration and definition are different from each other. He has recourse to a *reductio ad absurdum* to show that a definition does not prove: a definition 'proves either the "what it is", or what the name signifies' (92b26). But we have seen, Aristotle continues, that if it does not prove the 'what it is', then 'definition will be an account which means the same as a name'. But this is absurd for two reasons: first, there would be a definition of something which is not; second, any discourse would be a definition. Thus chapter 7 does not say at all that there is a definition – a nominal definition or any other kind of definition – of the goat-stag. Between chapter 7 and chapter 10 there is no variation in Aristotle's doctrine on this point.

Let us go back to the beginning of chapter 10. In no way can we find in this passage the idea that the nominal definition would be a kind of definition which would not presuppose the existence of the definiendum. The movement itself of the text shows this. We could paraphrase it in this way: one type of definition is what the term 'triangle' signifies. Having the knowledge that triangles exist (*echontes*, 93b32, is a present participle; Aristotle, therefore, does not want to say that this knowledge is acquired *after* the first one), we may go on and seek the cause. In other words, to have a nominal definition is to know what a term referring to really existing objects signifies. This is even clearer when we look at the summary at 93b38: 'one type of definition of definition is the one we have just stated. Another one is an account which shows why something is'. 'The one we have just stated' is the nominal definition. It seems that the nominal definition and the 'causal' definition are definitions of one and the same thing. But, as the second one is a real definition, that is a definition of a really existing thing, the first one cannot be only a signifier. We should probably not use any longer the phrase 'nominal definition', but, as I have nothing better to propose, I will follow this unfortunate tradition.

The problem of the number of the kinds of definition is, I think, both transformed and simplified by what has been said. If the so-called nominal definition is in fact a real definition, it would be bizarre that Aristotle should eliminate it from his final summary, as many commentators from

[25] 'The former type [of definition] means but does not prove, whereas the latter will clearly be like a demonstration' (93b39–94a2).

Philoponus onwards think he does. One may think, as Barnes does, that the list in the summary is not a complete one, but such a position has to be adopted only in a desperate situation, and it explains neither why the same list had already been given at 1.8.75b31, nor why the first type of definition should be eliminated. We therefore have to depart from most of the interpreters. The following is a paraphrase of the passage 93b38–94a10, which pays special attention to the connecting particles:

- On the one hand (*men*) the nominal definition is one kind of definition of definition; on the other hand (*de*) another kind shows why the thing is so. The difference between these two kinds is that on the one hand (*men*) the former one signifies, and on the other hand (*de*) the latter proves. But this proof is not a demonstration as it is usually understood but differs from it according to the position of the terms (the example of thunder follows).
- (Then comes a crucial passage.) There is also (*eti*) the definition of thunder as a noise in the clouds, which is the conclusion of the demonstration of what the thing is (by this, Aristotle means the 'causal' definition). I think that Ross is right to say that Aristotle does not allude here to a third kind of definition, but after he has dealt with the 'causal' definition, he just reminds his reader that there is a non-causal definition of the same thing, which is precisely the first kind listed, the nominal definition. That is why Ross puts into parenthesis the sentence at 94a7–9; as have I in my own translation.
- On the other hand (*de*) there is definition as a principle which is the posit of the essence without demonstration.

It is remarkable that the list of the different kinds of definition is encompassed within the *men*, *de*, *de* structure: the *eti* at 94a7 does not introduce a new kind of definition, as a *de* would do.[26] That this definition 'noise in the clouds' is the definition labelled as the nominal definition is quite clear: 'noise in the clouds' is what the term 'thunder' signifies. Ultimately, we actually find three kinds of definition in chapter 10: the nominal definition, the 'causal' definition and the definition as the posit of an essence. The definition which is the conclusion of the demonstration of the essence is another description of the nominal definition.

[26] It is interesting to notice that manuscripts d and n (which are among the best and the oldest) have *hoti* instead of *eti* at 94a7. The text would then mean: 'the same formula is presented in a different way, i.e. in one case we have a continuous demonstration, in another one a definition, *because* a definition of thunder is "noise in the clouds": this is the conclusion of the demonstration of the "what it is"'. Perhaps this is the original reading, which was corrected when people began to see in the 94a7–9 passage a third kind of definition. This reading is in any case in accordance with what Aristotle has in mind.

Such a reading of these chapters of the second book of the *Posterior Analytics* is far from solving all the problems of the sections of this treatise devoted to the theory of definition. We may, for instance, wonder to what extent the nominal definition and the definition as the conclusion of the demonstration of the essence are identical, if we accept the idea that they are identical. For there is a problem of priority between these two kinds of definition. The beginning of chapter 10 seems to describe a sequence which goes from the nominal definition to the 'causal' definition: we have an account of what the term 'eclipse' signifies – a loss of light of the moon – given that there are actually some eclipses; later we seek why it is so (cf. 93a23). But, on the other hand, if the nominal definition is the conclusion of the demonstration that proves that the moon loses its light because of the interposition of the earth, it is difficult to see how such a conclusion could be known prior to the demonstration.

Speculation on this point is quite hypothetical because of the obscurity of the texts and of the lack of examples. But we may get some information from the famous example of the eclipse as the inability of the moon to cast shadows (2.8.93a37ff.). I think that this example is not as artificial as scholars used to think it, but that it refers to plausible situations of people: hunters, sailors and the like, who may have to do something during the night and who may notice that objects do not cast the shadows they should, that the moon is not screened by clouds and that it is not a moonless night either, this last point requiring some astronomical skill. Therefore the moon itself does not have its usual light. Now, about this example, Aristotle says two things, namely that 'first, it is plain that the moon is eclipsed, but, second, not yet why, that is, we know that there is an eclipse but we do not know what it is' (93b2–3): we do not have the definition of the eclipse. Therefore grasping the inability of the moon to cast shadows is, according to the terminology used above (93a24), grasping 'something of the thing' non-accidentally. This means that, once we have grasped this, we can continue and seek the definition. But which definition? The passage 93b2–3 seems to mean the 'causal' definition: we do not know yet why there is an eclipse, that is (the *kai* at 93b2 is probably epexegetical), 'we don't know what [the eclipse] is'; we do not have a definition of it. Would Aristotle agree that 'inability to cast shadows' is a nominal definition of the eclipse?[27]

There is at least one reason which compels us to refuse to accept that the inability to cast shadows is a nominal definition of the eclipse, or if it is a

[27] The syllogism giving this nominal definition would be: 'Eclipse is the interposition of the earth, | Interposition of the earth prevents the casting of shadows, | Eclipse prevents the casting of shadows.'

nominal definition of the eclipse to consider that it is such in a derivative sense, namely that a definition, whether it is a 'nominal' one or not, has to be an account of the essence of the thing. But the inability to cast shadows is neither the essence of the eclipse, nor a part of it, but one of its *properties*. That is to say that such a character of the eclipse has to be *deduced* from what we know of the essence of the eclipse. Given that the essence of the eclipse is the loss of light of the moon, we can demonstrate, relying on the mechanism of the production of a shadow, that an eclipse entails the impossibility for the object situated under the moon to cast shadows. To say that the inability to cast shadows is the essence or belongs to the essence of the eclipse is to make a confusion between the level of the definition and level of the demonstration, whereas the very structure of Aristotelian science is based on their difference. We find again here the two levels I mentioned earlier about the different kinds of *per se* attribution.

The problem of priority or posteriority of the nominal definition which comes first in the list of chapter 10 relatively to the definition described as a conclusion of a demonstration remains so far unsolved.

We might, if not solve, at least reduce these difficulties in considering what is, according to Aristotle, the main goal of chapter 10 of *Posterior Analytics* 2. It seems to be to show that there are three types of definition, namely those which posit an essence, those which exhibit the cause of the definiendum and those which do not. In such a classification, the nominal definition and the definition as a conclusion of a demonstration are basically identical to each other, because both are definitions which do not mention the cause of the object. But what is even more interesting is to compare the 'causal' and the 'non-causal' definitions, e.g. the definition of the eclipse as a 'privation of the light of the moon' and as a 'privation of the light of the moon due to the interposition of the earth', or of thunder as a 'noise in the clouds' and as a 'noise in the clouds due to the extinction of fire' (94a5–9). It is clear that the 'causal' definition includes all the elements – all the information – which are included in the nominal definition, with something in addition. It is therefore hard to doubt that for Aristotle the 'causal' definition is superior to the other one. Or, in other words, the 'causal' definition is a condition for the existence of the non-causal definition, since the latter is the conclusion of a syllogism built up with the terms of the 'causal' definition put in a different order. That is, there would not be any definition, properly speaking, if the 'causal' definition and therefore the syllogism did not exist.

Such a distinction between 'causal' and 'non-causal' definitions is certainly to be related to the polemic against the Platonic method of definition by division. The impression that we get from the Aristotelian texts is that of a theoretical struggle between division and syllogism. Aristotle is far from being hostile to division, namely as a preparatory procedure for definitions. In this sense, one can even say that there would be no definition without division, but the method of division does not suffice to produce definitions.[28] Now Aristotle tries to protect syllogism, which he has presented at the very end of the *Sophistical Refutations* as his own discovery, from two sides: first, against division, in stressing that division does not demonstrate because it presupposes what it should demonstrate; but also against definition in showing that the definition remains within the sphere of essence, of which it is an account, whereas demonstration is concerned with the properties of an essence, and that it expresses these properties under the form of propositions, the predicate of which is a predicate related to its subject in a *per se* attribution, in one of the last two senses of '*per se*'.

The most surprising shift that we can notice about definition in chapters 8 to 10 of book 2 of the *Posterior Analytics*, is that after he has firmly underlined in the previous chapters that there are insurmountable differences between definition and demonstration, Aristotle now brings them very close to each other. As chapter 8 says in its final summary, if it is true that 'there is neither syllogism, nor, more precisely, demonstration of the "what it is", it is nevertheless through a syllogism and, more precisely, a demonstration that it becomes obvious' (93b16–18). Concerning chapter 10, as I have noted, one of its main goals is to show that there exist 'causal' and 'non-causal' definitions, the former being superior to the latter. Now, the definition which Platonic division is supposed to provide us with would clearly be a 'non-causal' one, and the Platonist cannot do better, precisely because he does not have the syllogism at his disposal.

To come back to the problem we are concerned with, namely that the nominal definition seems in one way to be prior and in another way posterior to the definition as conclusion of a demonstration, we may notice that this is a topic of difficulty in Aristotle, which he resolves by having recourse to his distinction between what is prior '*per se*' or 'by nature' and 'relatively to us'.

In the inductive procedure which builds up the definition starting from perception, it seems that Aristotle considers that we grasp a kind of nominal definition before grasping the 'causal' definition. The crucial passage is at 2.8.93a21–4 (see above, p. 136), a passage that we have to look at one more time:

[28] Cf. Bolton 1993.

As to whether something is, sometimes we grasp this accidentally, and sometimes by grasping something of the thing itself, for example of thunder that it is a sort of noise in the clouds, of an eclipse that it is a sort of privation of light, of man that he is a sort of animal, of soul that it is something which moves itself.

The preliminary stage is to know that the definiendum does exist. We have seen that Aristotle's position is that it is not sufficient to grasp this existence accidentally. I think that, at this point, the picture is not complete, because there are in fact three possible cases. If I grasp something accidentally related to the eclipse, that the dogs of the neighbourhood bark, for instance, this does not put me on the right path towards definition, because grasping this accident does not even establish that there is an eclipse. If I grasp the essence or something of the essence of the eclipse, namely that an eclipse is a sort of privation of light – Aristotle says 'a sort' because at this point I do not know the cause yet – I can proceed on the right path to the 'causal' definition. This is the way ancient commentators understood the 'something of the thing itself' at 93a22 (cf. Philoponus 367.31; Anonymus 562.28):[29] it is something of the *essence* of the thing that I must grasp to begin the process of grasping the 'causal' definition. But there is another case, which Aristotle considers further in the text, namely the situation in which I grasp a property, or, in Aristotle's jargon, a '*per se* accident', of the thing. Now we have seen that if I grasp a character such as 'the inability to cast shadows', which is a property of the eclipse, I grasp well and truly something of the thing in a non-accidental way, and therefore, I am on the right way leading to the definition.

I therefore think that there is no necessity that at the first stage 'something of the thing itself' indicates the essence, or a part of the essence of the thing. The only thing we are sure of, at the end of this first stage, which establishes the *existence* of the thing, is that there is an essence which can be defined. Such a certitude is not given by the barking of the dogs, but it is given by the inability to cast shadows. Actually there are two possible situations that can be described by the expression 'grasp something of the thing non-accidentally'. One can grasp a demonstrable property of the thing, such as the inability to cast shadows. We have seen above that, in this case, we do not have a nominal definition of the thing, but this does not prevent us from proceeding towards the definition. But one can also grasp the essence or something of the essence of the thing, for instance for the eclipse that it is 'a sort of privation of light'. In this case we have a definition of the thing – a definition being an account of the essence and not of a

[29] Wallies 1900.

property – and more precisely a nominal definition. On the other hand it is the case that such a nominal definition is formally identical with the conclusion of the demonstration the terms of which are present in the 'causal' definition.[30] It should be noted that here 'to grasp something of the thing itself' must be a grasping *per se*, but that this '*per se*' may have two senses, depending on whether it concerns the essence or the properties.

The nominal definition, for instance, for the eclipse, a sort of privation of light, when considered as preliminary to the grasping of the cause, is prior to the 'causal' definition relatively to us. But in such a role, the nominal definition can be replaced by a non-accidental property, that is a property which is *per se* in one of the last two senses of chapter 1.4. One may add that, when the nominal definition performs this function, it does not really have the role of a definition. For if what we grasp at this first stage can be expressed in a formula which is more or less identical with the nominal definition of the thing – 'for an eclipse that it is a sort of privation of light' in the sense that I grasp that it is the moon which loses its light – it is, so to speak, accidentally that we have grasped a definition of the thing. The text says explicitly that this first stage has to establish the *existence* of the definiendum, not its definition.

In the 'systematic' order, on the other hand, the nominal definition is posterior to the 'causal' definition, since it is the conclusion of a syllogism which is composed of the terms of the 'causal' definition, as Aristotle says explicitly at 94a8. And when one compares the 'causal' definition with the 'non-causal' definition, the nominal definition being of the latter kind, as I did above, we become aware that the only complete definition is the 'causal' definition, besides, of course, the definition as the posit of an essence. This is consistent with Aristotle's basic position at the beginning of book 2 of the *Posterior Analytics*, according to which the question of the essence and the question of the cause are one and the same question (2.2.90a31). Moreover it is remarkable that he repeats this position at the beginning of his 'scientific' inquiry, after he has finished with the 'diaporematic' one, on definition, as I have noted above: 'it is one and the same thing to know what something is and to know the cause of whether it is' (93a4).

From this, we can draw many conclusions. I would like to focus on two of them. The first one is that, though he keeps a strong distinction between demonstration and definition, Aristotle also establishes that it is only when you can demonstrate, that is, when you master the syllogism, that you

[30] This demonstration would be: 'Interposition of the earth produces a loss of light, | The moon suffers an interposition of the earth, | The moon suffers a loss of light.'

can really define. From this point of view, the real defeat of the Platonic division comes from the fact that it cannot play the role of the syllogism, as noted earlier. The second is that, all in all, the inquiry on the different kinds of definition ends with the 'causal' definition and the definition as the posit of an essence face to face. These are two types of definition which are irreducible to each other, because one presupposes the demonstration and the other is the basis for demonstrations.

TWO POINTS OF VIEW ON DEFINITION

The chapters dealing with definition in book 2 of the *Posterior Analytics* pay little attention to definition as a scientific principle, just mentioning it when the different senses of 'definition' are listed. On the other hand, these chapters insist on the definition as a 'demonstration of the essence'. And it is not difficult to understand why. The main problem with which chapters 2.3–10 are concerned is the relationship between demonstration and definition as *processes for knowledge*, whereas in book 1 the main question at stake in the texts dealing with definition is the *structure of scientific, namely demonstrative, knowledge*. Besides their differences in nature and function, these two approaches to definition are also characterized by a different relation to the basic Aristotelian cognitive procedure, the syllogism, and more precisely the scientific syllogism (demonstration). Within the framework of the deductive system that is demonstration, the principles, and first of all the proper principles that are definitions, have to be better known and prior relatively to what is deduced from them. Hence the dilemma which the last chapter of the *Posterior Analytics* intends to solve: how do we possess a knowledge 'more exact' (99b27) than the one given by demonstration of what has not been demonstrated? In such a schema, the demonstration is under the control of definition, since nothing is demonstrable but from true principles. But if we look at what definition gives us a knowledge of – on this point the unchanging Aristotelian doctrine is that definition provides us with the knowledge of the 'essence' – the relationship between definition and syllogism is quite different, since we have seen that the true definition is the definition which exhibits the cause; but for performing this function, definition needs the syllogism.

It would be inappropriate to say that there are two theories of definition in the *Posterior Analytics*. It would be particularly adventurous to say that the two approaches to definition we have detected – 'demonstration of the essence' and principle of science – belong to two different periods of Aristotle's research, given that this would go against the text itself.

For Aristotle feels no difficulty in putting side by side in the same list the 'causal' definition and the definition as a principle, for instance in chapter 2.10.[31] What these two conceptions of definition have in common, and which is an essential element of the structure of Aristotelian science itself, is that they are descriptions of *essences* of which sciences demonstrate properties. The fact that there are two different approaches to definition shows, among other things, that all the essences are not in a position of being principles. The common image of Aristotelian science as reflecting a real world constituted of ontologically autonomous essences (substances) is, then, basically right. Eclipse has an essence – even if, as Aristotle himself puts it, it is not the best possible example of an essence – which, of course, must be related to some more basic notions within the framework of 'great' sciences such as mathematics and physics; but it is also true that it has a reality of its own. That each of those essences has a definition is ascertained by the definition of definition as an 'account of the essence'. Each of these definitions is a source of knowledge in the sense that science demonstrates its properties, and this is precisely the basis of Aristotelian science. We have here a basic feature common to all the definitions, whether they are principles or 'causal' definitions, namely that definition is a starting point for demonstrations.

It would also be ill-advised to introduce a hierarchy between the different kinds of definition, in saying, for example, that the definition as an indemonstrable principle is 'more a definition' than the 'causal' definition. For each side of this double approach to definition in the *Posterior Analytics* may appeal to some superiority on the other side. In the demonstrative structure of science, the definition as a principle is a definition in an eminent way, but if we consider definition as a way of knowing, the 'causal' definition provides us with more information than the definition as a principle may do.

All this adds some weight to Brunschwig's argument: there is a part of the *Posterior Analytics* which presupposes a syllogistic, and a part which has to be related to a 'general theory of science'. The first part could be connected to the approach of definitions as principles, the second one with the construction of 'causal' definitions. Any further step would be pure speculation, but this does mean that we must not do it. I do not think we can attribute to these two approaches different dates, and Aristotle himself does not hesitate to put these two approaches side by side. I do not think either that science, as it is described in the *Posterior Analytics*, has to be

[31] But also in book 1, as I said above (cf. 1.8.75b31–2).

considered to be based upon a contradiction between two irreconcilable tendencies and that we may propose an additional version of an Aristotle torn apart between Platonism and anti-Platonism. It is clear, on the other hand, that for Aristotle to know in a scientific way supposes at least two distinct operations. The first one locates the knowable object within the structure of the science(s) of which it is an object; the second consists in grasping the essence of the thing and in deducing its properties from this essence.

CHAPTER 7

Male and female in Aristotle's Generation of Animals

Aryeh Kosman

> So God created the human in his image, in the image of God he created him, male and female he created them.
>
> (Genesis, 1.27)

> Urge and urge and urge,
> Always the procreant urge of the world.
> Out of the dimness opposite equals advance, always substance and increase, always sex,
> Always a knit of identity, always distinction, always a breed of life.
>
> (Walt Whitman, *Song of Myself*, 3)

I

Why is it that one animal is male and another female? This question, addressed by Aristotle throughout his treatise *On the Generation of Animals*, does not mean merely to inquire into the process by which a particular animal is determined to be one sex rather than another; that question, to which he turns late in his treatise, presupposes the larger and more radical question of why animals should be sexed in the first place. Aristotle provides the broad outline of an answer to this question early in his treatise: in animals that are sexually dimorphic, the male and the female are, he says, the *archai geneseōs* – the principles of generation, and specifically of animal reproduction, that is, generation of animals of the same kind.

I wrote an earlier version of this chapter for a 1983 summer workshop on Aristotle's biology organized in part by Allan Gotthelf. At the end of the summer I put it in a drawer and, except for one subsequent presentation when I received helpful comments from Sarah Broadie, did not look at it again. It seemed to me appropriate to resurrect it for a volume honoring Allan Gotthelf, partly in memory of how much I learned owing to him that summer. More importantly, my argument has been that we will not properly understand Aristotle's embryology unless we are clear about the ontology that underlies it and governs its articulation, and it has been Allan's vocation in his scholarly life constantly to attest to the importance and shape of that connection, to help us, as he has repeatedly done, to understand and to see the implications of the links between Aristotle's philosophical thought and his biology.

Animal reproduction, the generation by animals of others of the same kind – *anthrōpos anthrōpon gennōn*, for example, to paraphrase slightly Aristotle's description in the *Metaphysics*: human begetting human (*Metaph.* Z.7.1032a25) – is a good thing for animals, given that living is a good thing but that animals are mortal. Unable therefore to achieve the solitary eternity of an immortal individual, animals settle for and thus strive for the continuation of their species, that there might continue what is one not in number but in kind. Generation is this strategy of providing for the continuity of family given the perishability of family members, and it is for this reason that families go on though their members perish: *genos aei anthrōpōn kai zōōn esti kai phutōn*, as he puts it: the family of humans and of beasts and of plants is forever (*GA* 2.1.732a1–2).[1] The principles of this activity of generation, resulting in the existence of genera or families of animals, are the male and the female, and this fact answers one sense of the question "Why do male and female exist?"

Given the existence of a dual principle of animal generation, we should expect a differentiation in functional role for male and female; if there are two cooperative principles, there should be two different and coordinate functions. What are these different but coordinate functions? There is, according to one popular version of Aristotle's theory of animal generation, a simple answer to this question; according to that answer, male and female are distinguished by the fact that the male provides the *form* and the female the *matter* of the generated animal.

Such an account of what Aristotle takes to be the difference between male and female in the process of generation would, if true, have important consequences for how we view his larger understanding of the relationship between sexual differentiation and animal being in general. For if we think of an animal's form as the essence of what it is to be that animal, according to this account the male alone provides the animal's essential nature; the female provides merely the stuff from which an animal is made and therefore contributes nothing that is an essential part of the animal's being.

The male moreover is the active principle in the process of generation; the female is passive. The reason for this distinction is in turn that the male has a power that the female lacks; it is essentially the power of generation, and it is her lack of this power that leads Aristotle to describe the female as being "as it were, an *arren pepērōmenon*": a "mutilated male," as the phrase is standardly translated, or as A. L. Peck interestingly renders it

[1] See also *de An.* 2.4.415a28; for an earlier expression of this view, see Plato's *Smp.* 208a–b.

in the Loeb translation, a "deformed male" (*GA* 2.3.737a28).[2] So on this understanding of Aristotle's view, the female, itself not a fully realized animal, is an instrument essential to the process of animal generation but does not contribute anything essential to the animal that is the result of that process.

In this chapter I hope to show that this common reading of Aristotle, though understandable given some of his language, is seriously misleading. Aristotle's theory is not that the male provides the form and the female the matter of the generated animal, but that the male, in an act analogous to what we call fertilization, begins the process by which the female grows within herself and bears their mutual offspring. It is this process, the process of the generation of animals, in relation to which Aristotle speaks (though quite rarely) of the male as the formal and the female as the material principle, and in relation to which he identifies the male as active and the female as passive. To understand Aristotle's views on this subject, however, will require understanding correctly form and matter, and active and passive roles in relation to this process. Such understanding will I hope allow us to see more accurately Aristotle's theory of animal generation.

In its most extreme form, the view that I have suggested is commonly attributed to Aristotle can come to be assimilated to a view that he explicitly rejects in the course of the *Generation of Animals*. This is the view, which in its early modern form came to be called "preformationism," according to which the embryo or its parts are contained in miniature (in what some versions of preformationism called "homunculi") in the body of one or the other of the parents, most usually the father. In antiquity versions of this theory are found in Apollo's famous speech in the *Eumenides* of Aeschylus, in which the god allows that "the mother is not the parent of what is called her child, but the nurse of the fresh-sown embryo; it's the one who mounts that is the parent."[3]

Aristotle is clear and forceful in his rejection of the notion that the offspring itself must somehow be from the very beginning in the seed of one or the other of its parents (*GA* 1.21.729b1ff.). His rejection is made possible by a theory that employs the important concept of power or potentiality – *dunamis* – according to which the animal is contained in the seed only in the sense that the seed has the power to produce the animal by a process of what came to be called epigenesis. This paper is primarily a plea for the recognition that Aristotle's theory is through and through

[2] Peck 1963. [3] *Eumenides* 657–64; compare the speech of Orestes in Euripides' *Orestes* (552).

a theory of epigenesis, and an argument about what follows from that fact.

It is somewhat remarkable to see how often the preformationist theory is attributed to Aristotle; one author tells us that

> Aristotle buttresses his already strained teleology of female nature with a defective science of biological and reproductive processes . . . [holding that] . . . the male . . . implants the human form during mating. He deposits within the female a tiny homunculus for which the female serves as a vessel until this creature matures. The female herself provides nothing essential or determinative.[4]

Another indeed lays the origins of the preformationist theory directly at Aristotle's feet:

> It was not limits of observation that misled earlier centuries of scientists concerning the facts of generation; rather it was the 20-centuries-long concept, stemming from Aristotle, that women, as totally passive beings, contribute nothing but an incubator-womb to the developing fetus that springs full-blown, so to speak, from the head of the sperm. One's conceptual framework, a certain state of mind, permits one to see and accommodate certain things but not others.[5]

One indeed suspects that it may be a judgment concerning Aristotle's views on men and women, that is, his views on distinctions of gender, that permits this author "to see and accommodate certain things but not others" and that leads to such a radical misreading of Aristotle's views on sex and generation. In fact much recent literature about Aristotle's sexual biology is devoted to showing that it is his theory of the sexual differences manifest in generation that lead to his allegedly unsavory views on gender relations.

We may be able to understand what there is about Aristotle's views on men and women, or about views and institutions attributed to Aristotle, that might anger readers, but it is sad that this anger should lead to textual misreadings as radical as the ones I have just quoted. We need, in any case, to become clearer about what Aristotle does and does not say concerning the roles of male and female in the process of generation even to understand accurately what might be the implications of his views for what he says elsewhere about men and women.[6]

[4] Elshtain 1981, 44. [5] Bleier 1984, 3.

[6] Robert Mayhew (2004) provides a more extensive sampling of readings such as those I have quoted, and on a number of topics of which embryology is only one. Mayhew exposes with care and thoughtfulness how easy it is to misread Aristotle when other agenda are operative. He offers a treatment of the issues in animal generation that is somewhat different from what I here propose.

II

The view that according to Aristotle the female provides the matter of animal offspring is not a recent one. In his commentary on the *Metaphysics*, Alexander suggests that when, in a critical passage in book 9, Aristotle distinguishes between two analogous senses of activity or actuality (*energeia*), in one of which activity is "as motion in relation to a potentiality" and in another of which it "as substance in relation to some matter" (*Metaph.* Θ.6.1048b9), he means the analogy to be understood in the following way:

> as someone walking is to someone seated, so is Alexander to the menses. For someone is said to be walking as a motion in relation to a power, that is, as one attribute in relation to another, whereas Alexander is in relation to the menses as substance to matter; for the menses are the matter of a human being.[7]

Alexander's choice here of menses as the matter in question misunderstands and subverts Aristotle's argument.[8] For Aristotle has in mind not the menses, out of which an animal *comes to be*, but the living body, out of which, in the special sense demanded by his ontological theory, *it is constituted*. If there is any sense in which the menses are the matter for an animal and an animal may be said to come to be out of menses, it is not one in which the animal would be said to be, as it were, *menstrual*, as a box which comes to be out of the wood from which it is made might be said to be *wooden*; the animal, in other words, does not bear the same relationship to the menses that is its matter as it does to the body that is its matter. It must be, then, that the notion of matter is operating differently in the two different cases.[9]

We know from the discussions of the *Metaphysics* that, as a matter of fact, this is true, and that the two senses parallel the two senses of *dunamis* that Aristotle is so concerned to distinguish in book Θ of that work. Here is how we may think about it. A human body is qualified in several ways; among these modes of being qualified is the fact (of particular importance since it is essential) that it is alive in the way appropriate to it, that is, by being human. The relationship between the body of a human being and its being human – between a human body and the human being that it *is* – is a relationship that Aristotle describes in terms of his theory of *dunamis* and *energeia* – of powers and their modes of realization or exercise. In this

[7] Alexander, *In Aristotelis Metaphysica commentaria* 580.3ff. (Hayduck 1891).
[8] I give a more thorough argument for this view in Kosman 1984.
[9] I have set out to explain these different senses in Kosman 1984 and 1987.

instance the realization is a paradigmatic instance of that mode of *energeia* proper that Aristotle is careful to distinguish from motion. But the relation to the human body of menses as that which *becomes* the human body is one of realization *through motion*. This motion is the development of sperm and menses (or as we would now say, sperm and ovum) into the embryo and finally into the independent animal.

However mistaken Alexander may have been about the particular sense in which Aristotle thinks it true that menses constitute the matter of a human being, Aristotle apparently does think it true in some sense, and we may well wonder why even this should be so. Why should menses be thought to be the matter of the animal in any sense whatever? For thinking of there being matter that is then formed into an animal is to think of birth on the model of production, as though generation were a case of making something out of something. And this way of thinking seems to picture animals as though they were artifacts; it thus seems to introduce that very independence of matter and form in substance that the *Metaphysics* argues against.

Should we not instead expect to find Aristotle expressing a biological view of the coming into being of animals, a view that does not assimilate them to statues or wooden boxes? Such a view might be expressed most simply in the notion that an animal grows from seed. It does not come to be by a process of composition from independent matter and form any more than it is a composite of independent matter and form; its matter and form come into being simultaneously with one another and simultaneously with the complex entity that is the animal. Why then does Aristotle describe the process of generation in terms of matter conveyed by the female and form by the male?

III

One possible answer that has proved attractive invokes the feature of animal reproduction that I cited at the very beginning of this essay. Aristotle is interested in the fact that, among complex animals, natural birth is as homospecific as it is heterosexual, and perhaps these two features explain one another. It is true and important that in the *Generation of Animals* Aristotle is interested in the homospecificity of animal birth and is concerned not merely, as we might carelessly think, with one specific instance of the general problem of being arising from non-being, not merely with "proliferation and ontogeny." If cats gave birth indiscriminately to different entities – kittens, puppies, chicks, or tufts of grass – the general problem of ontogeny would remain. But Aristotle's interests are more particular; they are tied to one of the deepest and most remarkable features of animal life and therefore of the being of substance: the fact, with which we began, that animals replicate themselves in kind, that cats have kittens and dogs puppies. It is, as we might say, not merely genesis but genetics that interests Aristotle in the *Generation of Animals*: not simply how generation takes place, but why it takes place so as to guarantee resemblance between generator and generated. More particularly, it is the resemblance of kind that concerns him here and in general, the phenomenon of homospecificity; the fact, as he puts it in the treatise *On Generation and Corruption*,

> that from a human being always or generally comes a human being, and from wheat wheat, and never an olive. (*GC* 2.6.333b8–9)

Here is an answer some have taken to be Aristotle's; it is an obvious answer, but its details are what are important, and if we take those details as part of Aristotle's embryology, they might serve to explain his theory of sexual function. Parents and their offspring are homospecific because a parent passes its specific form on to the offspring. We might then go on to say – and here perhaps we take Aristotle to be influenced by Greek culture in general and by his own views on the relative natures of male and female – that it is the male parent who does this, who passes on through his sperm the form of human being to the offspring. If we do not slip into thinking of Aristotle as a preformationist, we can recognize that he thinks of this passing on as purely formal; the father's contribution is in no way a material contribution. The sperm does not form any part of the *body* of the fetus or new offspring, but only contributes to it its *form*: human being, horse, or whatever. The body, on the other hand, is, on this reading of Aristotle's views, the contribution of the mother, who provides through her menses the material out of which the fetus is formed. So the whole picture is one that is startlingly modern in its appreciation of the way in which genetic transfer is a formal transfer, and not surprisingly sexist in its view of the relative roles of male and female. The mother supplies the matter and is thus the ultimate source of the body, the father supplies the form and is thus the ultimate source of the soul, and all this is offered by way of explanation of the homospecificity of parent and child. Such a theory would provide a means of explaining sexual dimorphism in terms of homospecificity. It is this view of Aristotle's theory that I mean to call into question.

IV

I suggested that if Aristotle's theory of animal generation depicts the father as providing the form of an animal and the mother its matter, it is a theory inconsonant with his larger metaphysical views. Note that this would be true even if animals were not substances; for even the creation of an artifact does not take place by a joining together of form and matter, but by a process of forming the matter. What is especially true in the case of substances is that there is no constitutive matter that exists independently of the substance, and this fact increases the general oddness of supposing that this is an accurate view of the generation of animals. Indeed, the simplest objection to thinking of anything as the matter of a human being in this sense (and therefore to thinking of one thing as contributing matter and another form) is the fact that such a vision of how animals come to be turns out to be a vision of how animals are made. And animals, as I have maintained, are not made; they grow from seed. To imagine that they are made is to think of taking matter and imposing a form upon it, as one might imagine (think of the creation myth in Genesis 2) a god picking up clay and breathing animal life into it.

But indeed, none of this is Aristotle's view. To begin with, he does not claim in the *Generation of Animals* that the sperm of the male parent contains the specific form of the adult animal in any sense in which the female menses does not. There is, of course, a trivial sense in which human sperm may be said to "contain" the form of human being; for it is a part of the human being and so has as part of its formal definition the form of human being, as does eye or blood or marrow. But in this sense so do menses; what is not true is that human sperm "contains" the form of human being in any special sense that explains the formation of human fetuses and the birth of human babies.

In fact, with very few exceptions, Aristotle does not characteristically describe the male as providing the form of the animal. What he says instead is that the sperm constitutes the first mover of a process that results in the adult animal; it is as *archē tēs kinēseōs* – a first principle or origin of motion – that the male and the sperm are first and typically described (*GA* 1.21.730a27; 2.1.734b23; 2.4.740b25). The sperm is a residue whose essential nature is in the power that it possesses by virtue of its motion, a tool that nature uses as something actively exhibiting the relevant motion – *echonti kinēsin energeia* (*GA* 1.22.730b21). As such, it initiates, by that motion, the series of motions that result in the existence of an individual having the form of the adult animal; but it does not itself in any special sense *contain* that form.

The view that animals become formed by a successive series of motions that take place in the mother is at the heart of Aristotle's epigenetic theory, and the claim that that series begins with the motion imparted by the sperm is what connects the sperm to that epigenetic theory. To suppose that the sperm operates by containing the substantial form of the animal in question is exactly to forsake epigenesis in favor of a theory that is essentially preformationist, even if it is what we might call formal rather than material preformationism. Most simply, this study, as I earlier remarked, is a plea for recognition of Aristotle's *formal epigenesis*, a reminder that on Aristotle's view the form or soul of the generated animal must come into being gradually during embryonic development, just as its body does. The form of the animal is not fully present at any time during the motion that is embryonic development, except potentially: not as the *eidos* or form of a very small animal, and not as a disembodied program or code for an animal.

If it were indeed true that the male is responsible for contributing the specific form of the generated animal, then it would be true that the female contributes nothing essential to the determination of that animal, and there would indeed be little difference between Aristotle's view and the outlandish view of Aeschylus' Apollo. The standard difficulties of interpreting Aristotle's discussion of resemblance and inheritance in *Generation of Animals* 4.3 would appear moot; for it would be a mystery why Aristotle should have thought that the mother needed to be of the same species as the father, or indeed why there should be any such phenomenon as hybrid animals. The mule, a favorite animal of Aristotle's, would be a totally inexplicable phenomenon. Furthermore, Aristotle's concern to explain the necessity of the male in generation (*GA* 2.5.741a6) would be strange; we should expect him to be concerned to explain the necessity of the *female* in the process of generation.

But these difficulties disappear once we are clear that the essential function of the male is to initiate a series of motions that take place in the mother, remembering that these motions continue in an automatic epigenetic fashion and that the menses remain throughout part of the mother's body, so that the development is affected both by the nature of the menses (different for different animals as any other animal part is) and by the continuing motions of the mother which constitute her life and support of the embryo. As Aristotle says:

> By the embryo, I mean the first union of male and female. (*GA* 1.20.728b34)

V

I said that, with few exceptions, Aristotle does not describe the male as providing the form of the animal. But what of those exceptions? Does not Aristotle indeed talk about the male as possessing and contributing to generation *form*, even at the same time that he characterizes the sperm as *archē tēs kinēseōs*, the principle of motion? The male for instance is said to

> furnish the form and the source of motion (*parechetai to te eidos kai tēn archēn tēs kinēseōs*), while the female furnishes the body and the matter (*to sōma kai tēn hulēn*). (*GA* 1.20.729a9–11)

More generally, in the very midst of his account of generation, Aristotle characterizes as better and more divine the

> first cause of motion (*archē tēs kinēseōs*) to which belong *logos* and *eidos* – definition and form. (*GA* 2.1.732a3–4)

He cites this as a reason for the separation of male from female. So even if the male or the male's sperm is the first cause of motion, the *archē tēs kinēseōs*, it would seem also to be that to which belongs *logos* and form.

What could Aristotle mean by describing the male as that "to which belong *logos* and *eidos*"? It is difficult to believe that he might by this phrase mean that the male itself has the logos and form of *human being*, for example, or horse or dog, whereas the female does not; for no human being is more or less a human being than any other, no horse more or less a horse than any other. Male and female are not thought by Aristotle to differ in relation to *what* they are, but merely as different modes of the same essential nature, somewhat as people of different complexions differ from one another.

To be sure, the situation is more complex than that; the difference between the sexes is precisely of interest because it is not *merely* accidental, but functional. Neither, however, is it essential, that is to say, it is not the expression of a difference in species. It may then be of larger metaphysical interest to ask whether there are other such differences, and how they fit into Aristotle's ontology.[10] Clues to how such a question could be answered may be found in the brief and puzzling remarks at *Metaph.* Z.5.1030b16–27 on the differences between the sense in which an animal is white and the sense in which an animal is male or female. But whatever we say, women and men, like mares and stallions, remain members of the same *species* and must therefore both exhibit the essential form of what it is to be human or equine.

Nor does it seem likely that Aristotle means to suggest that the *sperm* contains the *logos* and *eidos* of the animal, that is, the animal's specific form, except in the qualified sense that any part or organ of the animal does. For, in the first place, the sperm itself would then be not a part of the animal, but something like, *per impossible*, the very animal itself with a severely truncated body: that is, a homunculus. And there is the point; Aristotle would then, as I suggested above, be espousing the very preformationism that it is his intention to avoid.

What then could he mean? Here I want to detour for a moment and look at the language of some contemporary accounts of animal generation, accounts that I think may help us to understand what Aristotle is doing. Here is a description of sexual reproduction from what was some years ago a standard university biology textbook:

> Sexual reproduction involves two parents, each of which contributes a specialized cell, a gamete (eggs and sperm) which fuse to form the zygote, or fertilized egg.[11]

Note here the move from thinking of the zygote as the fusion of ovum and sperm, the result of a yoking (as its name suggests) of two gametes, to thinking of it as a fertilized egg, as an ovum that has been fertilized, the result of the activity of one gamete upon another. This move is a common one, and both descriptions are found in biological writing side by side, as though there were no essential distinction between the two, as though what resulted from the fusion of an egg and a sperm were a fertilized egg. The authors of this textbook go on to say that the typical life cycle of an animal is a simple one of growth of a fertilized egg into an adult which produces more eggs for fertilization,[12] and this way of speaking is carried on in clear view of the mechanics of genetic contribution, which they describe in these words:

[10] The complexity of this issue is brought out in Marguerite Deslauriers's thoughtful essay "Sex and Essence in Aristotle's Metaphysics and Biology" (Deslauriers 1998).

[11] Villee and Dethier 1971, 815.

[12] Villee and Dethier 1971, 822.

The union of one haploid set of chromosomes from the sperm with another haploid set from the egg, which occurs in fertilization, reestablishes the diploid chromosome number. Thus the fertilized egg or zygote, and all the body cells which develop from it by mitosis, have the diploid number of chromosomes.[13]

Here is another description from a classic textbook, this time by no less a biologist than James D. Watson:

New organisms arise from the highly differentiated sperm and egg, whose union initiates a new cycle of division and differentiation. Thus, although a complicated organism like man contains a large number of cells . . . all these cells arise initially from a single cell. The fertilized egg contains all the information necessary for the growth and development of an adult plant or animal.[14]

I have quoted these modern authors because it seems to me that their ways of speaking, and in particular their characterization of a zygote as a fertilized egg, the result of an egg's having been fertilized by a sperm, would be quite familiar to Aristotle, and because seeing this may help us understand what Aristotle's meaning is. Note first that to think of the zygote as a *fertilized egg* is to think of the egg as *passive* and the sperm as *active*. By "passive" I do not here intend the secondary sense of *torpid* or *idle*, nor the further derivative sense of *submissive*; I mean only to translate Aristotle's *pathētikon*, and so to refer to that which is *acted upon* in contrast to that which *acts*. The understanding that associates passivity with idleness comes from the conflation of two different though related senses of *activity*. To avoid confusion that might result from this conflation, we need to keep in mind that acting is one characteristic form of *energeia* – activity – and being acted upon is another; in both forms a subject is active in the one sense, but in the first case it acts and is in a different sense active; in the other it is acted upon and is in that sense passive.

Describing the egg captured by the phrase "fertilized egg" as passive is then in a sense merely to remark upon the grammatical form of "fertilized" as a passive participle. But of course not simply grammatical: when we think of the formation of the embryo as a process of fertilization, we are thinking of the male gamete as doing something to the female gamete, namely fertilizing it, and of the female gamete as having something done to it by the male gamete, namely being fertilized. The act of sexual reproduction is thereby figured as beginning with this process of action and passion, namely fertilization, a process that results in what we call the fertilized egg, and in which the fundamental roles of male and female are those of acting

[13] Villee and Dethier 1971, 232. [14] Watson 1970, 10–11.

and being acted upon. This process is exactly analogous to what Aristotle describes when he says that the male *qua* male is active and the female *qua* female passive in that process by which the semen *sets* the menses to form the embryo, where *set menses* is analogous to *fertilized egg*. As Aristotle puts it:

> *To ge thēlu hē thēlu pathētikon, to d'arren hē arren poiētikon kai hothen hē archē tēs kinēseōs*: the female *qua* female is passive [i.e. is acted upon]; the male *qua* male is active, that is, is the initial source of the motion. (*GA* 1.21.729b12–14)

To call the male active and the female passive (which is only slightly less deceptive than calling the male *poetic* and the female *pathetic*) may further mislead us unless we remember also that for Aristotle to be passive characteristically involves a capacity or ability that an entity possesses; it may in other words represent a *dunamis* in the strongest sense of that word. We need only think of the many powerful *dunameis* that Aristotle thinks of as passive: the senses, the emotions, the mind itself. It is important in this respect to keep clearly in mind that the passive is not, in this sense, "just passive."

To think of the zygote or the embryo as a fertilized egg or as set menses is, secondly, to think of the egg or the menses as matter and the sperm (in both our sense and Aristotle's) as the agent by virtue of which we may say *matter* is *formed*. But the matter and form in this saying are not the matter and form of the *adult animal*; they are rather the matter and form of the *zygote or embryo*, that is, of the *fertilized egg or set menses*, in so far as it constitutes zygote or embryo.

This point is less a specific biological one than a general metaphysical one. Let me put it schematically: if some (bit of) M has the further characteristic of having been F-ed or of being F, and by virtue of this fact is an (instance of) E, then M and F are respectively the matter and form of the entity E; the entity just is formed matter, or matter in a form. If, for instance, a piece of bronze, by virtue of having been shaped into a statue or being statue-shaped, comes to constitute or constitutes a bronze statue, then bronze is the matter, and statuesque shape is the form of the statue; or if an oak beam, by virtue of having been placed in or being in a certain position, constitutes a lintel, then the oak beam is the matter, and the position the form, of that lintel; for the statue is shaped bronze, and the lintel is a positioned beam. We will, of course, need to be able to tell a story about the relationship between the different senses of *being formed* that we have just invoked, and that story will necessitate our understanding the relationship between becoming and being. But having told that story, we will be able to say, with Aristotle, that just as the bronze is matter for the statue and the beam matter for the lintel, so, if an egg or menses by virtue of being fertilized or set comes to constitute a zygote or embryo, then the egg or menses is the matter, and the fertilizing or setting sperm provides the form of the fertilized egg or the set menses. Recall the interesting language of the first of the textbooks we listened to: "an adult which produces more eggs for fertilization."[15]

On this reading, we can begin to see how natural it is to speak of the female as the source of the matter and the male as that of the form without privileging one or the other with respect to the essential features of the generated animal. Recall in this respect how easily Aristotle himself moves from speaking of the fetus as the union of male and female (*migma thēleos kai arrenos*) to speaking of the male as providing the form and the female the matter (*GA* 1.20.728b34–729a12).

VI

Why do biologists (and the rest of us as well) speak of the zygote as a fertilized egg, rather than, for example, a fertilized sperm, or perhaps an egged sperm? Why do we think of the sperm as doing something to the egg, and not the other way around? Any answer to this question will have to be somewhat speculative, but we can make some guesses. I shall begin with two that come from biologists to whom I have posed the question in conversation. (1) It might be, one colleague suggested, that the egg, relative to the sperm, is large, and, as she put it, sluggish and indolent, whereas the sperm is small, and in her words, motile and aggressive. It may seem unlikely that Aristotle, or anyone whose embryology is developed at the macroscopic level, might have entertained this notion; but if we think of avian eggs, which Aristotle studied intensively, it might seem less unlikely.[16] (2) It might be, another colleague suggested, that the egg, as it were, stays home while the sperm goes abroad, and that the place of the zygote is therefore the place of the egg. Generation thus takes place "with" the egg by means of the agency of the fertilizing sperm; precisely because the egg cannot become the embryo except through the agency of the sperm, it is the (fertilized) egg that becomes the embryo.

These suggestions are, as I said, highly speculative; they are meant merely to remind us that many people, eminent modern biologists among them, speak about generation in ways compatible with Aristotle's embryology.

[15] Villee and Dethier 1971, 822. [16] See *HA* 6.1–3.558b7ff.

Given his conceptual categories and philosophical vocabulary, someone could easily agree with Aristotle that "the male generates in another, the female in itself," and that "the male is active and the female passive," and that "the male provides the form and the female the matter of the embryo," without in any way denying the real and significant genetic contribution of the female or the female's formative role in reproduction.

It may be precisely the feature of generation remarked on in the second suggestion that Aristotle has in mind when early in book 1 he distinguishes the coordinate functional roles of male and female. Male and female differ, he there says, because their powers differ one from the other,

> the male being that which has the power to generate *in another* and the female that which has the power to generate *in itself* and out of which comes to be a generated that exists *in* the generator. (*GA* 1.2.716a20–3)

Notice that thus to describe the male and female as distinguished by possessing different but coordinate powers is to attribute to each a power that the other lacks. The male lacks the power to generate in itself, a power that the female possesses, just as the female lacks the male's power to generate in another. This way of putting things, however, is somewhat disingenuous, for it is only the latter inability, the female's inability to generate in another, that Aristotle notes and upon which he remarks, just as though the only significant power were that of the male. The question I have just posed – "why speak of a fertilized egg?" – is precisely the question "Why, given the existence of coordinate powers in the male and female, does Aristotle emphasize the ability of the male to initiate the process of embryonic growth?" Why is the male, so to speak, thought to generate the process of generation?

But even to recognize the male's power as one of a coordinate pair of powers may be helpful. For one thing, it should help us to understand a notorious and generally misunderstood remark in book 2. Aristotle has just observed that the female's bodily contribution to generation, like male semen, is also a residue and also contains all the parts of the animal body potentially, though none actually, including the parts that distinguish the male from the female. He then proposes an explanation of how the female can contain potentially, that is, be able to produce, the sexual parts of a male. To this end, he offers his readers a figurative analogy:

> Just as the offspring of disabled parents (*pepērōmenōn*) are sometimes disabled and sometimes not, so the offspring of a female is sometimes female, sometimes not but rather male. For the female is, as it were, a disabled male: *to gar thēlu hōsper arren esti pepērōmenon.* (*GA* 2.3.737a25–8)

To figure the female as a "disabled" male is not to describe the female as mutilated or deformed; it is to register by strong metaphor the fact that the one sex lacks an ability that the other has. A female is, in this trope, comparable to, as it were, a castrated male, an animal that lacks the power of initiating generation, that is, as we have seen, of fertilizing. It does not lack anything essential to the nature of the animal in question, but only a power specific to the male. The meaning of the figure is revealed in an earlier version of Aristotle's remark:

> a woman is, as it were, an infertile male – *estin hē gunē hōsper arren agonon* – for it is on account of a certain inability that the female is female. (*GA* 1.20.728a18–19)

It is, to be sure, the female that is figured as lacking, and so we need to ask, as we have been asking, why that is the case. The female has a corresponding power, the power to generate within; but we find no passage that figures the male as a hysterectomized female. And so the question I have posed concerning our tendency to speak of a fertilized egg presents itself in yet another form: why does Aristotle give prominence in the course of his explanation of generation to the male's power to initiate the process of generation in another?

Perhaps thinking about our puzzle in terms of one power's being explanatorily privileged over the other will be helpful. Think back to how Aristotle opens his treatise on the generation of animals. He notes that his discussion so far, presumably in *On the Parts of Animals*, has dealt with the causes of animal life in all relevant senses of cause but one; it has considered formal, final, and material causes, but not "the *archē tēs kinēseōs*" the "principle of motion," that is to say, the cause responsible for initiating the process by which animals come into being. In this work, he says, he will give a description of the parts of animals devoted to generation, and consider the "*aitia hē kinousa*," the "motive cause." And, he goes on to say,

> inquiry into this [motive cause] and inquiry into the generation of each animal is in a sense the very same thing. (*GA* 1.1.715a14–15)

These remarks suggest that Aristotle views the *Generation of Animals* as a work devoted not simply to explaining the conditions of generation, but specifically as well to explaining the *process* of generation, and in particular the initiation of that process. Generation in this respect is explained by appeal to the *archē tēs kinēseōs*, the cause that is responsible for generation's getting underway. Aristotle's theory identifies the power of the male as precisely that power to initiate the process of generation. So this understanding of the explanatory project of Aristotle's treatise on generation provides a different answer to the question why he attends specifically to the power of the male. It is the male, most simply, that has the power to start things going, to initiate and thus in this genetic causal sense to bring about the generation of a new animal.

VII

Besides attempting generally to domesticate Aristotle's language of matter and form, action and passion, these last remarks have had a more specific purpose; it is to emphasize again that when Aristotle speaks of matter and form in the *Generation of Animals* he generally understands not the matter and form of the animal that is generating and being generated, but the matter and form of the entity that is in process of being generated or of the process itself, that is, the matter and form of the embryo.

What on this view is the form of which Aristotle is speaking? It is first the form of what for him is the analogue of being fertilized, namely the motion by which the menses first becomes the developing embryo. In this sense, as I have suggested, the male clearly does provide the form, since the male parent provides the semen which contains the motions that transform the menses into fertilized menses. It is secondly the form of the continuing process that results in the animal offspring, the entity in motion that is the developing embryo. Embryos, of course, do not just happen to become the animals that they become; it is no accident that a human fetus becomes a human being. This is because such an embryo is a human being in the making. And so for embryos in general; it is essential to their nature to become the animals they are in the course of becoming.

If what I have suggested has plausibility, it may make Aristotle's views on generation seem less outrageous. Should we be concerned that it does so at the expense of making unclear the very features of homospecificity in generation that I earlier invoked as a possible explanation of why Aristotle speaks of matter and form? For if it is correct that Aristotle rarely mentions form in his account of generation, and if when he does so he is referring to the form of the embryo *qua* embryo and not to the specific form of the parent (male or female) *qua* the kind of animal it is, then it would seem that the power of the theory as an explanatory device for homospecificity has vanished. If the form of the animal is not passed on, what does explain the fact that *anthrōpos anthrōpon genna* – human begets human, and not horse or hippopotamus or mink?

But my contention has not been that animal form is not present in the act of generation; it must be, as Aristotle insists repeatedly, present in the parent(s). But it is not, I have meant to argue, present in the sperm (except in the trivial sense mentioned above) nor is it conveyed in the sperm; Aristotle's view is that animal form is passed on by means of the sperm and is reproduced through the agency of the sperm but is not transmitted *in* the sperm (nor *a fortiori in* the menses). Just so, in Aristotle's analogy, the form of the table may be present (in some sense) in the mind of the carpenter and may be transmitted to the table by the active kinetic agency of the tools. But it is never present *in* the tools, except of course, *dunamei*, as a *power* that the tools have, the power to form a table from wood. And this is what Aristotle means by speaking of the animal existing potentially – *dunamei* – in sperm and menses alike.

We may put the matter this way: the passing on of animal form from parent to offspring is not an explanatory principle in the *Generation of Animals*; it is (one of) the items that he sets out to explain, and he explains it not by further reference to a form passed on like my grandfather's pocket watch, but by reference to a complex theory of spermatic motions, menses, pneuma, animal heat, and so on, a theory meant to explain how form is passed on. The sperm initiates a series of motions that eventuate in the birth of a young animal by virtue of its powers as sperm. But these powers are not more uniquely associated with the specific form of the animal than are those of any other functioning part of the animal – blood, marrow, or, at a higher level, eye and liver. Because the sperm's function is to produce another living animal of the same species, it is said, by virtue of these powers, to contain potentially the soul, that is, the specific form of the animal, as it contains potentially the body and its parts, and exactly as the female menses are said to do, presumably by virtue of their *dunamis*. But the sperm does not contain the form in any *actual* sense; it operates as a mover, that is, as an initiator of a process that results in there being another individual of such and such a form. It is this power to *initiate* the process of generation that the male possesses and the female lacks, and it is this power with respect to which the female is said to be impotent in comparison with the male. But this asymmetry of power with respect to initiation is not true of generation *per se*. In general the sexes are distinguished primarily, as we have seen, not by the fact that one has power and the other lacks it, but by the fact that they have different powers; one is capable of generating in itself and one capable of generating in another.

These two powers are made active only when the male initiates the set of motions that constitutes generation. It is in the context of the power to effect that initiation that Aristotle speaks of the female in terms of privation. It is also in the context of that power that a female offspring is said to be in a sense a prodigy, deviating not from being human, but from having the power of the parent who initiates her conception. If the initiating movement of the male is dominant, Aristotle writes,

> it will produce a male and not a female, and a male who resembles its male generator and not its mother.

But if it is not dominant (we might say recessive)

> then in whatever power it fails to be dominant, it will make the offspring lack that power. (*GA* 4.3.767b21–3)

If the sperm is essentially a first mover or the *archē* of a *kinēsis*, then an account of its nature will have to involve reference to this *kinēsis*. We may want to say that the embryo itself is this motion,[17] or, as I suspect is preferable, we may want to say that the embryo is an entity whose nature is to be in this motion, the motion initiated by the sperm in the menses and eventuating in a new animal. In either case, we will need to remember that a motion, like any other entity, has a form, and that its form will necessarily involve its whence and whither, what it is a motion toward (recall the discussion of *EN* 10.4.1174a29–b8). The form of the sperm is thus not that by virtue of which it is the initiator of just any old motion, but of a very particular motion of a very particular sort – a motion which, when it takes place not just any old where, but in matter of the right *dunamis*, namely the menses of the right sort of female animal, will result in not just any old offspring, but in an animal of a specific nature: a horse, a hawk, a human being, or a hippopotamus. One animal initiates the generation in another animal of yet a third animal all of the same kind not by communicating to that animal a specific form, but by initiating in that animal a motion of a determinate nature that will result in an animal of that specific form.

VIII

I have argued (1) that form is rarely mentioned in the *Generation of Animals* in descriptions of the male or sperm, (2) that the male and its sperm are rather described as the initiator of the generative motion, (3) that when form is mentioned it is characteristically the form of this motion or of the embryo that is meant, and (4) that this makes better sense of Aristotle's theory of kinetic epigenesis. Homospecificity and family inheritance are therefore explained in a fashion that I would describe (cautiously and with appropriate qualifications) as "Aristotelian" rather than "Platonic" or say

[17] This is the suggestion of Cynthia Freeland in her essay, "Aristotle on Bodies, Matter, and Potentiality" (Freeland 1987).

rather as parental rather than patriarchal: I mean, without a leap of form from father to child.

I want to suggest finally that these questions of homospecificity and the subsidiary forms of resemblance between parent and child are not the only questions that interest Aristotle in this work, and they may not even be the most central; it is not just the mechanics of generation and inheritance that attract his attention in the *Generation of Animals*, but the question as well of why generation employs a differentiation of the species into two sexes, and, more generally, why it should be the case at all that certain kinds of living organisms exhibit such a differentiation.

The *Generation of Animals* opens with a description of what Aristotle terms *to thēlu* and *to arren* – The Female and The Male, whose neuter forms, as Peck points out, indicate that the discussion is about female and male as principles, principles, as we immediately learn, of generation in animals.[18] Book 2 then begins, as noted at the opening of this study, with an inquiry into the reason for there being male and female. The account that immediately follows, in which male and female are explained by appeal to an explanation of that generation of which they are the principles, is only one half of the account; it gives the reason in terms of a functional teleology of sexual differentiation. What it does not give is the account that we might expect given Aristotle's theory of hypothetical necessity: an explanation of why, given that end, this should be the means for its achievement. Why it is that animals of a certain level of complexity employ both males and females in the process of reproduction?

This question concerns the principles and causes of sexual differentiation, not in the scientifically interesting but relatively minor sense of book 4, which asks how individuals come to be determined as one gender or another, but in terms of the larger question of why there is, so to speak, the institution of gender at all. Nor can it be answered simply in terms of the question why generation demands two parents; it must be framed in terms of the question why generation demands two different kinds of parents. It is not uncommon to encounter in contemporary discussions of this question accounts of the origin of sexuality in terms of genetic recombination. But while such accounts might explain isogamous multiple parentage, they do not explain the central fact of sexual dimorphism; they do not explain, in other words, why the male and female gametes are different in nature from one another.

[18] Peck 1963, 10, note a.

But it is this question of why the male and female principles are different and separate that lies behind much of the discussion of the *Generation of Animals* and lies at the heart of Aristotle's concern with the different roles of male and female in animal generation and, as he hints early on, "*en tō holō*," which Peck cleverly renders in English as "in cosmology" (*GA* 1.1.716a15).

I do not know whether Aristotle ever answers this question. I have argued that the fundamental difference in his view between male and female is not between two modes of animal being, one of which is fully human and the other somewhat less than human, nor between two modes one of which is human and the other merely a deformed human, nor between two modes one of which has the power of generation that the other lacks. It is rather the difference between two modes of animal being that exhibit complementary powers of generation. The male has the power to engender, the female the power to bear. Aristotle, as we have seen, expresses these complementary powers of originating and of giving birth by describing the male as that which is capable of generating in another, the female as that which is capable of generating in itself. I have argued that this characterization is directly linked to his view of the male as the *archē tēs kinēseōs*, the initiator of the motions that lead to birth.

It is precisely this description that Aristotle invokes when, at the beginning of book 2 of the *Generation of Animals*, he argues that nature has kept the male separate from the female, that is, created different sexes, because the male is better since, he says, it is the *archē tēs kinēseōs tois ginomenois*: the initiator of the motions that result in generated animals (*GA* 2.1.732a7–9).

This description in turn may lead us to conjecture that Aristotle's understanding of the differences between men and women has to do not with some notion of being more or less human or more or less rational, but with what he takes to be their relative capacity to act, as it were, at a distance and as an instigator, to initiate and govern productive activity without necessarily participating in it. One might say: to rule. This is a topic for another discussion, but I mention it in closing, admittedly with chutzpah, as a protreptic. A feminist perspective in reading Aristotle will perhaps best be served not by charges of misogyny, nor of Aristotle's holding a view of woman as not fully human, nor for that matter of his doing bad science. We need instead to begin by reading carefully Aristotle's actual science; we can then consider, for example, the implications of his characterization of the male as *archē tēs kinēseōs*, a characterization that I have argued is central to his views concerning sex in the *Generation of Animals*.

CHAPTER 8

Metaphysics Θ.7 *and 8*

Some issues concerning actuality and potentiality

David Charles

INTRODUCTION

In *Metaphysics* H.6 Aristotle famously remarks:

What then is the cause of what is potentially F being actually F in the case of things that come to be over and above the efficient cause? For, nothing other is the cause of what is potentially a sphere being actually a sphere; rather this [i.e. the cause] is what it is to be for each of them singly. (1045a30–3)[1]

This passage and its immediate context can be interpreted in several ways. One interpretation, which I have defended elsewhere, runs as follows.

> 'If we consider the issue of the unity of a composite in terms of
> matter : form
> potentiality : actuality[2]
> there is no longer a difficulty. What makes it the case that
> potentiality : actuality
> are paired in such a way as to form a composite unity is that they share a cause: what it is for each of them to be what they are (a formal cause).'[3]

An early version of this chapter was read at the École Normale Supérieure in Paris in January 2004 and I am indebted to Justin Broackes, Michel Crubellier, Mary Louise Gill and David Lefebvre for their comments on that occasion. It was revised and extended for a conference held in honour of Allan Gotthelf in Pittsburgh in October 2004, where I gained greatly from his comments and those of Jim Lennox. I have subsequently been helped by discussions with Ursula Coope, Edward Hussey, Michael Peramatzis and Dory Scaltsas. My attempt to connect issues in *Metaphysics* Θ.7 and 8 with Aristotle's teleology stems from an understanding, secured by many enlightening conversations with Allan Gotthelf, of its central role in Aristotle's metaphysics. In this chapter I have sought to characterize Aristotle's teleology in a way which abstracts from issues we have discussed elsewhere. Indeed, my aim has been to sketch an account of Θ.7 and 8 with which Allan could, in general terms, agree.

[1] In this translation 'this' is taken to refer back to 'cause' in 1045a32. Other possible translations are mentioned below.

[2] In this summary, 'potentiality' and 'actuality' are simply used as stand-ins for '*dunamis*' and '*energeia*'. I shall use 'capacity' and 'potentiality' interchangeably in this essay. Others prefer 'activity' as a translation for '*energeia*'.

[3] I have argued for this interpretation in Charles 2000, 294–300.

Consider an example: in the case of man the actuality, according to this view, is *being alive in a given reason-involving way*. This is what it is to be a man: the relevant formal cause. The matter in question is made what it is by this formal cause: it is what is capable of being alive in this way. When what is capable of being alive in this way is actually alive, there is a unified composite. The unity in question is non-accidental: for the matter is defined as what is capable of being alive in a given way, the form as being alive in that way. Aristotle's re-conceptualization of the unity of the composite in terms of potentiality and actuality enables him to show how the issue of its unity is solvable. The *relata*, matter and form, are non-accidentally connected in the way potentiality and actuality are.

H.6, it must be admitted, leaves many problems unresolved. It says little about what the formal cause is, other than noting that it makes the composite a unity and an F. Although in H.4 Aristotle had remarked that the final and the formal causes are often the same and in Z.17 had pointed to the final cause as the formal cause in the case of things that are (presumably including substances), he does not refer back to these remarks in H.6. Nor does he give an account of the actuality (*energeia*) or capacity (potentiality: *dunamis*) involved; still less does he show how the former is prior to the latter. If Aristotle is attempting to clarify the unity of the composite by introducing these terms, he needs to address these issues. One would expect this to be his project in *Metaphysics* Θ. Is it?

Aristotle's strategy in *Metaphysics* Θ appears relatively clear: he begins by examining a basic sense of 'capacity' (*dunamis*) and 'actuality' (*energeia*) in which the active power to produce a change in something else is called a 'capacity' (1045b34–1046a4) and the change it produces (an activity) is called 'an actuality' (*Metaph.* Θ.1–5). He then turns to talk about applications of these terms to further cases. We learn that the capacity to undergo a change can also be called a 'capacity' (1048a29) as can (*inter alia*) (a) the ability to think or see (1048a33; 1048b1f) and (b) matter. The latter are describable as capacities by analogy. Thus, Aristotle notes:

> matter stands to *ousia*
>
> as
>
> [the] capacity to produce a change stands to the corresponding change

He remarks, replacing talk of 'changes' with talk of 'processes':

> All things are not said to exist actually in the same way, but only by analogy . . . for the relation is either that of process (*kinēsis*) to capacity (*dunamis*) or substance (*ousia*) to some type of matter . . . (1048b8–9)

The questions raised in H.6 now seem pressing. Aristotle needs to say how matter (of a certain type) can be correctly described as regarded as 'a capacity', substance (*ousia*) as 'actuality' and the relation between them be described as analogous to that between the capacity for a process and the process itself. He also should set out the advantages of applying the terms 'capacity' and 'actuality' to the case of substance. He must do both if he is to provide a reflective understanding of his use of these terms in this context. My aim in this chapter is to examine how far and how successfully he achieves this goal in Θ.7 and parts of Θ.8.[4]

The line of interpretation just sketched remains controversial. One problem is this: in H.6 talk of capacity (*dunamis*) and actuality (*energeia*) was, as has just been suggested, introduced to illuminate the relation between matter and form in a composite substance. But in 1048b8–9 it might look as if Aristotle is using these terms to describe the relation between matter and the composite substance itself. For in Θ.6 the substance in question is, it appears, not the form but rather the composite: the finished substance, the statue of Hermes (1048b3f.). If so, it might look as if in Θ.6 the unity of the composite substance is simply taken for granted, not explained (as in H.6) in terms of actuality and capacity. Has Aristotle quietly given up in Θ.6–8 his attempt to articulate the unity of the composite substance in terms of its more basic components?

Reflection on the last point might support an alternative, non-explanatory, reading of Aristotle's strategy in H.6 and Θ.1–8. According to this view, the key notions of actuality and capacity are best seen as abstractions from a more basic notion of one unified substance. According to this alternative interpretation, the unity of the composite is taken for granted (as an actuality) and matter understood simply as what is potentially a unified composite.[5] There is, therefore, a challenge for those of us who favour the explanatory interpretation sketched above: to find in *Metaphysics* Θ.6–8 answers to the following questions:

1. What is meant by conceiving of matter as capacity (*dunamis*)? What are the advantages of so doing?
2. What is meant by conceiving of substance or form as actuality (*energeia*)? What are the advantages of so doing?
3. What is the basis for the analogy between capacities for change and capacities (or potentialities) for substances?

[4] These chapters raise many other fascinating issues which lie outside the scope of the present study. Some will be mentioned *en passant*, others simply ignored.

[5] Those who have favoured this interpretation include Wilfred Sellars: see Sellars, 1967, 73–124. I sketch their view in Charles 1994, 87–93. For further discussion of this issue, see Lewis 1994.

I shall examine what Θ.7 (and especially 1048b37–1049a18) and Θ.8 (especially 1050a4–b5) contribute to these issues. My proposal is that these sections contain answers to these three questions, ones which rest in large measure on Aristotle's interest in teleological explanation. I shall first offer a reading of parts of Θ.7 and (from p. 183) of certain lines in Θ.8. I shall then sketch (on pp. 183–97) Aristotle's answers to the questions just outlined, commenting on their significance for his account of matter and substance.

METAPHYSICS 1048B37–1049A12: THE BASIC MODEL

1048b36–1049a2: what question is being raised in this chapter?

We should determine when each thing is potentially [something] and when not. For it is not potentially something at any time whatsoever. For example: is earth potentially a man? Or not, but rather when it has already become the sperm or not even then perhaps. Just as not everything is capable of becoming healthy (whether through medical skill or by chance), but there is something which is capable of becoming healthy and this is what is potentially healthy.

Here is one way to understand this question:

(A) 'Under what conditions is A potentially B?'

(A) can be understood as the traditional question: 'what conditions are necessary and sufficient for A to be potentially B?'

Here is another:

(B) 'At what point/time is A potentially B'?

(B) can be understood as follows: in a sequence of a given kind, what is the first point at which it is true to say 'A is potentially B'? This will be the first point at which A is potentially B; before that it was not potentially B.

The differences between questions (A) and (B) are important. Here is an example of question (B):

At what point is something potentially an oak tree?

One answer to this question is:

When it is an acorn and can grow into an oak.

Before that stage there was nothing which was potentially an oak tree. Two points should be noted about this answer:

1. Possessing the qualities involved in being an acorn is not sufficient for the presence of the finished tree. The tree has yet to develop. The acorn may not get there.
2. In the process of becoming an oak tree, the acorn will grow, develop and change. Its original abilities and capacities will be refined, modified and

altered in the process of the growth of the tree. Although, throughout the process, the object will retain the capacity to be an oak tree, what is involved in having that capacity will change over time.

Several factors in *Metaphysics* Θ.7 suggest that Aristotle is seeking to answer question (B).

1. The immediate text contrasts the relevant time (or point) with others: some before the crucial time *(oupō*, 1049a17), some just after it (*ēdē*, 1049a15; see *Ph.* 222b7f.), anytime you like (1049a1). The original question is posed as follows: 'At what point in the following sequence is something potentially a man: earth, sperm, some later stage (1049a1–2)?' At some stages it is not then (*tote*, 1049a2) a man. What is being sought is the point in this temporal sequence at which it is true to say 'A is potentially a B' such that before then it is not true to say this.
2. Later, Aristotle focuses on the issue of where in an ascending series something is rightly called potentially a B (earth, wood . . . : 1049b19ff.). Here, too the question seems to be: 'At what point in the ascending series is something potentially a box such that before then it is not potentially a box?'
3. Conversely, Aristotle nowhere explicitly addresses the further issues required to answer question (A): (i) How long does A continue to be potentially a B? (ii) At what point does something cease to be potentially a B (when it is a B, on B's death, on B's decomposition)? For an answer to question (A) requires one to say what are continuity and exit conditions of the relevant potentiality.

If one takes Aristotle as answering question (A) in Θ.7, one might suggest that he offers an implicit answer to this question: A is potentially a B for just as long as it retains the potentiality to become a B. But he fails to offer any reason for taking the potential to be a B to be coterminous with the potential to become a B. Nor is it at all obvious that these two potentials can be coterminous. Consider the case of a fully formed (or ageing) human being: its matter is still, it seems, potentially a human being even though it is no longer capable of becoming a human being. At the very least, much would need to be said to argue the contrary and there is no sign of the required argument in *Metaphysics* Θ.7–8. Indeed, if we assume (in the light of Aristotle's remarks elsewhere) that what is actually a B is also potentially a B, it looks as if something can be potentially a B and lack the ability to become a B.[6] If so, it seems safest to take Aristotle as focusing in Θ.7 not on the full range of conditions under which something is potentially a B but

[6] For an alternative view, see Frede 1994, 188–93.

on the narrower issue of the first point at which something is potentially a B.

So what is Aristotle's answer to question (B)?

1049a5–12: the external cause

The dividing point (limit) in the case of what comes to be in actuality [*entelecheiai*] from what is potentially in production by practical thought is when it comes to be (actually) at the will of the agent without any hindrance from factors outside his (or her) will, and in this case (the one just mentioned) when nothing in the patient being healed hinders his (or her) being healed. Similarly, in the case of what is potentially a house. If nothing in it (i.e. in the matter) prevents it from becoming a house, and there is nothing which needs to be added or taken away or changed (for it to become a house), it is potentially a house. And the same account applies in all cases where the starting point of the process of coming to be is external (to the matter).

What is the point at which something is potentially a house? Aristotle is clear: the point at which nothing in the matter prevents its being turned into a house. All that is required is the decision (and subsequent actions) of the builder. But which point is this? Several possibilities have been canvassed.[7] I shall consider three: (1)–(3).

(1) The matter is potentially a house at that point at which it can be turned immediately into a house (and not before). Nothing needs to be added, taken away or changed for the house to be completed.

But this suggestion is problematic: strictly understood, it would mean that something was only potentially a house when all the bricks except the last one had been put in place. It would only be at the very last step in the process of building that something was potentially a house. However, so understood, it would not be bricks and pieces of wood that are potentially a house but rather bricks and wood arranged *in a given way* (in the all but finished house). And this does not seem to be the way in which Aristotle conceives of his examples (see 1049a19ff. and the embryo case below).

(2) The matter is potentially a house at that point at which no further material change is required for it to be ready to be turned into a house by one activity of an art (house building/medicine) and not before.

[7] Several of these were discussed at a reading party on Θ.7 held at the European Cultural Centre in Delphi in 2002. I am indebted to those who participated in those discussions and to two major recent interpretations of this chapter (Gill 1989 and Frede 1994). I have not attempted to specify in detail where I agree or disagree with either Gill or Frede, still less to argue systematically for my interpretation as against theirs. My aim is only to offer an alternative reading of Θ.7.

The idea in (2) is that the matter is potentially a house at that point at which it could be handed over to the builder for him (or her) to make a house by the exercise of his (or her) skill as a builder (and not before).[8] The relevant exercise of the skill will be one which begins with bricks in a certain state and turns them into a house. Before there were bricks, another skill was needed (other than that of the builder) to produce matter on which the builder's skill could operate.

This proposal is problematic in two respects.

First, it seems to lack the generality required for Aristotle's purposes. For something can be potentially a B not only when it can be turned into a B by one act of an art but also when it can be turned into a B by some internal natural cause (1049a11ff.). Reference to skill is not wide-ranging enough for present purposes. Nor is it clear how it is to be generalized.

Second, it is difficult to see how talk of skill can provide the reflective understanding we require. For it immediately raises the following question: how are the operations of the relevant skill themselves to be individuated? Presumably, these are processes and as such individuated by their starting and finishing points (see *EN* 1174b5). The finishing point is the completed object. But what sets the starting point of the operation? What is it about bricks that make them (rather than cement) the appropriate starting point for the relevant skill of house building (let alone one operation of it)? Why is there one skill (let alone one operation of a skill) which takes one from bricks to a house? Why not take there to be one skill (or even one operation of that skill) which begins with cement (or cement mix and water) and ends up with houses? Perhaps one can find one craftsman who can do both. Why should we not say that he has just one skill: that of making earth into houses?

To address these two issues, one needs to see skill as a special case of some more general phenomenon and to explain why the starting points for operations of one skill are to be found at one point rather than another. We can address the first challenge as follows:

(3) The matter is potentially a house at that point at which no further material change is required in it for it to be ready to be turned into a house by an actuality-inducing agent and not before.

This proposal differs from the second in attempting to fix the lower boundary at which the relevant art is to begin by specifying the point at

[8] This suggestion requires that the matter in question could be handed over to the builder, not that it is actually handed over. Although the builder might be several hundred miles away waiting for the matter to arrive, the matter for the house would exist when it left the factory gates. I am indebted to Larry Jost for advice on this point.

which the matter is ready for the *actuality-inducing agent* to do its work: nothing need be added or taken away from the matter for it to be ready for the actuality-inducing agent to begin the process of turning it into the relevant actuality. Skill is an external, nature an internal actuality-inducing agent.

But what of the second question: what is to count as an actuality-inducing agent? Why does it, in the case in point, start its work with bricks rather than cement? A sentence in *Metaphysics* Θ.8 provides a clue:

> Further, the matter is potentially [an F] when it would proceed towards the form; when it is actually [an F] it is in the form [enformed]. (1050a15–16)

Something is potentially an F when it will (if all goes well) turn into an F.[9] I shall begin with an overview of the relevant context in Θ.8 and seek to provide a basic model of what is going on. I shall fill in some of the relevant background in Θ.8 below (pp. 183–7).

The sentence just quoted is part of a passage in which Aristotle argues for the priority in being of actuality (*energeia*) over capacity (*dunamis*) in composite substances (1050a4–b4). His argument (1050a7–23) is heavily dependent on teleological concepts: goal (*telos*, 1050a8, 17 18, 21), that for the sake of which (*hou heneka*, 1050a7), in order to (*hina*, 1050a1–2). At the outset, Aristotle reasons as follows:

(1) In cases of coming to be, everything proceeds towards its goal (1050a7–8): premise
(2) [In these cases] the goal is the relevant actuality (*energeia*): (1050a9): premise
(3) In cases of coming to be, everything proceeds towards the relevant actuality (from (1) and (2))
(4) The relevant capacity (*dunamis*) is present for the sake of it: premise
and concludes
(5) The actuality (*energeia*) is prior to the capacity in being (1050a4).

(1) introduces the idea of the goal and a general teleological principle: in these cases (if all goes well) everything proceeds towards its goal. (4), it appears, extends the operation of the teleological cause back beyond the stages of the process to the capacity for the change. It too is present for the sake of the goal and will (if all goes well) assist in producing the goal.

Aristotle next gives a range of examples designed to support (4): we have sight in order that we may see; we possess the capacity to build in order

[9] For this reading see Burnyeat 1984, 142.

that we may build; we possess the capacity for theoretical activity in order that we can engage in such activity (1050a10–15). This teleological order is (in some way) taken to underwrite the final conclusion (5).[10]

In 1050a15–16 (quoted above), Aristotle is seeking to apply the line of thought he has just developed to the case of matter and composite substance.

The first claim:

> Again matter is potentially something when it will (if all goes well) proceed towards the form

is, it appears, an instance of premise (1) above, with 'matter' replacing 'everything' and 'form' replacing 'goal'. Aristotle certainly intends the form to be a goal. For he continues by identifying actuality (*energeia)* with the goal in the case of processes and of nature (1050a16–19). The object (or person) achieves the relevant goal when they are active in the appropriate way; for the goal is to be active in that way. He explains this as follows:

> For the result (in the case of action) is a goal and the actuality (*energeia*) is the result. (1050a21–2)[11]

The result (*ergon*) in cases of action can be either a finished product (e.g. an enformed statue) or the activity itself (such as flute playing).[12] From these claims it follow that the actuality in question is a goal. In some cases, the goal is the finished product (1050a25–7, 30–4), in others it is to be active in a given way (1050a23–5; 1050a34–b1).

[10] How to understand (5) and the associated claim of priority of actuality over potentiality? While an examination of this question lies outside the scope of the present study, I should note that I do not understand the claim to require that the relevant actuality can *exist* without its associated potentiality. The actuality may be prior *in being* to the potentiality in that it is what it is independently of the potentiality being what it is (although the converse is not true). This order of being is the basis for the order of definition. So understood, the line may be prior in being to the triangle (even if both are necessary existents). I am indebted at this point to many discussions with Michael Peramatzis; see Peramatzis 2008, 187ff.

[11] In cases involving matter, the form (being an F) is the goal both of the process of becoming an F (1050a8–10) and the potentiality to be an F: the form is the actuality (*energeia*) in question. Further, in this case the actuality (*energeia*) is an *entelecheia:* the completed or perfected result which is to be achieved as the goal in question (1050a21ff.). Aristotle notes that the term '*energeia*' is complex and can apply both to actions (such as using knowledge) and to the perfected state (e.g. of a statue: 1050a20). While in both cases the goal is an actuality (*energeia*), in the first it is an activity (in the case of (e.g.) seeing) and in the second the perfected state which results from the process of making a statue (or building a house). '*Energeia*' can refer both to actions and to the results of actions. I discuss the act/result ambiguity of '*energeia*' in Charles, 1986, 132–9.

[12] Both could be called 'works': art works (such as statues) and the work of the flute player (e.g. flute playing). For discussion of this ambiguity, see *EE* 1219a12–17. Given these considerations, '*ergon*', like '*energeia*', is act/result ambiguous.

Aristotle, it seems, is already committed (in 1050a15–16) to the following claims:

[A] The goal for matter is to be enformed.

[B] To be enformed is an actuality (*energeia*, in the sense of *entelecheia*).

From [A] and [B] he can infer:

[C] The goal for matter is an actuality (being enformed).

If so, he can redeploy the original argument (1050a5–15) in the case of matter and form as follows:

(1)* In cases involving matter, matter will proceed towards its goal (being enformed) (premise)

(2)* The goal is an actuality (premise)

(3)* In cases involving matter, matter proceeds towards its relevant actuality (from 1 and 2)

If he can secure as a further premise:

(4)* The relevant capacity (*dunamis*) is present for the sake of the goal,

he will be as well placed as before to conclude:

(5)* The actuality (*energeia*) is prior to the capacity in being (1050a4).[13]

Further, since in these cases the form is the relevant actuality (1050b2), Aristotle can also infer (after an interlude focusing on somewhat different points (1050a23–b2) that

(6) The form is prior to the capacity in being (1050b3ff.).

This discussion helps to fix more precisely the idea of an actuality-inducing agent. We can now see the importance of the term '*entelecheia*' in 1049a5–6: what is salient about skill is that it brings about a goal of this type, a perfected state, by thought. Skilled actions are explained by the goal which the agent seeks to achieve. So too is the presence of his (or her) capacity to build, since this too is acquired for the sake of this goal. In these cases, the goal in question teleologically explains both the actions that lead to it and the presence (and nature) of the required capacity. An actuality-inducing agent is an agent (i) whose actions (and capacities) are teleologically explained by the relevant goal in this way and (ii) who (if all goes well) achieves that goal.

I shall try to spell out this idea in more detail, using Aristotle's favoured example: house-building. The form of the house guides the action of the builder who turns the bricks into a house but it does not guide the actions of the craftsman who turns earth into bricks; for the actions of the builder are

[13] There remains a problem as to how (5)* follows from the premises: see also n. 10 above. It is left unaddressed here.

sensitive to the fact that the goal in question is a house. It is the presence of this goal that explains why he does some things and not others. His actions, beginning with the bricks, are intelligible only on the assumption that he is building a house (and nothing else). By contrast, the form of the house does not guide the actions of the craftsman who is turning earth into bricks; for his actions would have been just the same whether the final product was a house or (e.g.) a town wall. His actions are not made intelligible by the fact that a house is the final product. (And this would be so even if he was in fact aware that he was contributing to the building of a house; for even then his specific actions would have been made intelligible by the goal of building a house. He would have acted in just the same way had he been contributing to building a town wall!) Intelligibility by reference to the relevant goal is lost once one goes below the level of the bricks (in the case of house building). This is why the relevant skill (house building) begins with bricks and not with earth. And this will be so even if one man has, as it happens, both skills and can begin a set of operations with earth and end up with the finished house. For his actions will not be guided throughout in the favoured teleological way by the final goal of the house he builds.

On this understanding, the matter is potentially a house at that point at which no further material change is required for it to be ready for the form to be imposed by an agent (i) whose actions are intelligible in the way just specified by the relevant goal (the form) and (ii) who will (if all goes well) achieve the relevant goal. It will be in such a state when and only when that agent's subsequent actions can be explained (teleologically) by the fact that it is this form (and no other) that is being imposed. If so, the potentiality to be an F will extend back as far as do the actions of such an agent which are teleologically sensitive (in the way just sketched) to the fact that it is this form that is his (or her) goal. The potentiality to be an F is first found at the point at which the intelligible action of this agent commences.

Does this proposal have the necessary generality? Let us turn to the second case discussed in *Metaphysics* Θ.7: 1049a14–18.

THE BASIC MODEL EXTENDED: 1049A13–24

1049a13–18: the internal cause

In cases where there is an internal starting point of the coming to be, A is potentially a B when a B will come into being by itself. For example, the sperm is not yet potentially a man (for it needs be placed in something else and change); rather something is potentially a man when it is the state in which it can be a man by virtue of a starting point which belongs to it.[14] The sperm needs another starting point to turn into a man in just the same way as earth is not yet potentially a statue (for when it changes it will be bronze).

In the case of an external starting point, the matter is potentially a house at that point at which no further material change in matter is required for the form to be imposed from outside (by an external agent) governed by the relevant teleological end (in the way just specified). If the analogy holds, in the case where the starting point is internal, A will be potentially a B at that point at which A is in the state required for the relevant form to be imposed from within (if nothing prevents) by the operation of a principle internal to A (governed by the relevant teleological end).

Let us spell out the details more fully. The sperm lacks the type of matter required for the operation of an internal principle whose actions are governed specifically by the goal of becoming a human being. It is not yet in a state ready for the relevant actuality-inducing agent to do its work. For it to be ready it has to change (1049a15). By contrast, the *embryo* will be potentially a human at just the stage at which no further material change is required for its development to be governed solely by a cause internal to it, one guided by the goal of developing into a human being. In this it is like the bronze on which the statue maker operates: nothing needs to be added to it to make it be in the state required for his actions to be governed solely by the goal of forming a statue. Indeed, Aristotle adds this example presumably to show what is common to the cases of internal and external change. Art and nature are both governed by similar teleological principles (1049a16–18).

The example of the sperm sharpens our understanding of Aristotle's claim. Its actions (and nature) can, no doubt, be explained teleologically by the goal of making a human being. However, it is not itself capable of being a human being since it is not in a state in which the relevant internal agent can operate on it so as to make it into a human being. What is important for something's status as matter is its being in a state on which the relevant actuality-inducing agent can operate. While the actions of the sperm may be sensitive to the goal to be produced, it is not what produces the finished product. The sperm needs to be changed before it is ready for the operation of the teleologically sensitive agent whose actions culminate

[14] Punctuating the Greek text with a colon, not a comma, after *metabalein* (1049a15) to allow Aristotle to distinguish between sperm and embryo.

(if all goes well) in the relevant goal. For this reason the sperm is not the matter of the human being.[15]

There is another aspect of this example which requires comment: the embryo, when not separate from the mother, has the nutritive soul only potentially (*GA* 736b13ff.). As such, it is potentially but not yet actually a human being (as the wood is potentially but not yet actually a house). Similarly, the *embryo* is potentially a human body and so potentially something which is necessarily alive in the ways a human is but not yet itself alive in the way a human is. Of course, for Aristotle, the human body, once formed, is both potentially and actually alive as a human (*de An.* 412a20f.). Indeed, it seems that a human body is necessarily alive as a human. If so, there are two separable ways of being potentially a human. The *embryo* is potentially but not yet actually alive as a human being while the human body is both potentially and actually alive as a human being. In the latter case, the matter is actually an F: something which is enformed by F (1050a15f.). While both body and *embryo* can be described as potentially a human being (and as the matter for a human being), the ways in which they are so differ. Indeed, the *embryo* offers an example of something which is potentially (alive as) a human being but not yet actually (alive as) a human being. Something can first be potentially a B when it is capable of being turned into a B in a given way and later be potentially a B when it has in fact been turned into a B.

To make the last point quite specific: in varying stages of development, there will be something which is capable of being a man in that it is capable of being organized by the form of a man. Something will be capable of being a man if either (a) it is in state ready for the operation of an agent specifically sensitive to this goal or (b) it is in a state which is currently

[15] At what point does the *katamēnia* itself have the potentiality to be a human? Is it only when it has been fertilized or before? The issue presumably is this: is the *katamēnia* ready for the internal principle to operate before it is fertilized or only when it has been fertilized? Aristotle is not clear on this point here. There are two possibilities. First, if one focuses on the analogy with bricks and the builder, the *katamēnia* (prior to fertilization) will be in a state ready for the internal principle to operate. It is ready for the internal agent to do its work. Alternatively, one might take it to be required not only that the *katamēnia* be ready for the internal principle to operate but also that the internal agent be present and actually operating. In which case, the unfertilized *katamēnia* would not yet be potentially a human. Which is correct? In the present context, one might translate 1049a15–16 as saying it is only when it is in a state in which it can be operated on by the relevant internal principle that it is potentially an F (in line with the first alternative), or as saying it is only when it is being operated on by the relevant internal principle that it is potentially an F (in line with the second).

It may well be that Aristotle is not (at this point) concerned to adjudicate between these two possibilities. This said, Aristotle's analogy with the case of the bricks and the builder seems to favour the first alternative: the bricks are potentially a house not only in the presence of the actual builder but also when they are (e.g.) on the way to the building site.

organized so as to realize that goal. In the first case it will only be potentially a man; in the latter it will also be actually a man. The matter of a man is what is capable of being a man in either of these ways. There can be considerable changes in what constitutes the matter in question provided that throughout these changes it is capable (in one of these ways) of being a man. In the later stages of the story, the matter of the human will be his (or her) body. In the earlier, the matter will be the *embryo* capable of developing into a man under the operation of the relevant internal agent. Both of these will be what is potentially a man (*to dunamei F*) provided that both (in their differing ways) possess the potentiality (or capacity) to be a man.[16]

1049a18–24: more of the same?

It seems that when we describe something not as a this but by the '. . . *en*' locution (as we call a box not wood but wooden and wood not earth but earthen and similarly with earth if in a similar way it is not a this but describable with the '. . . *en*' locution), that [which we use to call something else '. . . *en* . . . '] is always, without qualification, potentially the next thing up in the series. For example, the box is not earthy nor earth but wooden. For wood is potentially a box and the matter of a box . . . wood in general of box in general and this wood of this box.

This section follows the pattern of thought in the preceding sections. Aristotle remarks that we call something (e.g. a box) 'wooden' when the latter (the wood) is potentially the former. It is natural to interpret this in the light of his remarks earlier in the chapter as follows:

> The wood is potentially a box at that point at which no further material change is required for the form to be imposed by an external agent (i) whose actions are intelligible in the way just specified by the relevant goal (the form) and (ii) will (if all goes well) produce the finished product.

By contrast, earth is not potentially a box since further material change is required for the form to be imposed by the external agent whose actions are intelligible only in the light of the goal of the finished product: the box. The nature of the transition from earth to wood is not itself sensitive to

[16] On this understanding, the matter of a human being can exist (and be conceived as existing) other than in an enformed human body. For it can pre-exist the formation of such a body. This is consistent with accepting that the body of a human being cannot exist (or be conceived of existing) except as in an enformed human body. For the matter of a human being can pre-exist the formation of the body in question. This point is relevant to one aspect of a famous puzzle raised by Ackrill 1972/3.

the fact that it is a box rather than (e.g.) a table which is being produced. In this way, Aristotle has principled grounds for blocking the following inference:

> A [earth] is potentially B [wood], B [wood] is potentially C [a box]
> so A [wood] is potentially C [a box].

For while A is intelligibly connected (in the way specified) with B and B is also intelligibly connected with C, A is not intelligibly connected in the required way with C. Once one gets below the level of B, the transitions that occur to generate B are no longer intelligible on the basis of what is required for *a box* to exist. Rather, they are intelligible in terms of what is required for wood to exist. There is no single thread of intelligibility which leads from the top level to the bottom.

The direction of intelligibility is important: what is to count as the matter of the box is determined in terms of what has to be present if the form is to be imposed in the teleologically based way just specified. Thus, for example, what it is to be the relevant matter of the box is determined by what is required for the form to be imposed by the relevant agent in the way just specified. There is no way of specifying what the matter is except in these terms. If one tries to do so, one loses one's understanding of why this matter has to be present if the box is to exist. The relation is asymmetric: given the presence of the wood one cannot specify which form will be realized (as the wood has many potentialities). One could, of course, specify which form will be realized if one adds reference to the actions of a carpenter who aims to transform the wood into a box. But this will only be because he is guided in his actions by the relevant form (of a box).

There are several further points to make about this passage:

1. The wood (this wood) is undetermined (1049b1) only in the sense that it has not as yet been made into the (new) determinate box by the imposition of the form of the box. At this stage the wood has several potentialities (to be a box, to be a table, etc.), and it is only when the form of the box is imposed that it is determined which of its potentialities will be realized. In this way, the form of the box acts as a determinant selecting which of the potentialities of the wood is realized in forming a new determinate object: the box.
2. If wood is that from which a box is formed and is also potentially a box, there is no difficulty here in allowing that the matter from which something is formed persists in the composite. It persists in that one of its potentialities is exercised. For the introduction of talk of matter as the potentiality to be a B (understood as above) allows matter to be both

that from which the box is formed and to persist in the box. For at both stages its potentiality to be a box is present.

3. On this understanding, the analogy with the universal is precise but limited: neither (this) wood nor whiteness is yet the determinate thing in question (this box: this white object). Both require a determinant to make the determinate object in question.[17] There is no requirement that the wood in the wooden object itself becomes indefinite having originally being something definite (this wood).[18] All that is needed is that one of its potentialities persists and is actualized in the box. It does not follow from this that matter is like a universal in any further way: being indefinite while in the new composite, and so forth etc. And this may prove a benefit: the presence of one persisting potentiality is just what is required to allow for the new composite not to be a *de novo* replacement of the original matter but something in which that matter persists.

SOME SECTIONS OF Θ.8 AND THE BASIS FOR THE ANALOGY BETWEEN PROCESSES AND SUBSTANCES

It might reasonably be objected, at this point, that more needs to be said in favour of the teleological understanding of the notion of matter in Θ.7 just sketched. My interpretation has rested on (i) giving the term '*entelecheia*' in 1049a5–6 a teleological gloss (what has the goal in it) and (ii) pointing to some teleological material in Θ.8 to account for one use of matter as potentially a B. But, it will be said, I need to show first that the teleological material in Θ.8 is important for that chapter and, second, that it is at work in Θ.7 (or before). I shall seek to address these concerns in the next two subsections.

Teleology in Θ.8

In *Metaphysics* Θ.8, as noted above, Aristotle commits himself to the following claims concerning the case of matter and form:

[A] The goal for matter is to be enformed.

[B] To be enformed is an actuality *(energeia)* (1050a15–16).

From [A] and [B] he can infer:

[C] The goal for matter is an actuality (being enformed).

[17] At the bottom of the series, fire is not a determinate object at all but just stuff, since it is not itself made determinate by a determinant: 1049a24–6.

[18] For a contrasting view, see Gill 1989, 158ff.

[C] is an instance of a more general claim:

[C]* The goal for matter and for processes is an actuality (see 1050a9).

Aristotle tends to reify talk of actuality and form in this passage and permit himself to draw the further conclusion that

[D] the form is an actuality (1050b2–3)

and finally that

[E] the actuality is prior in being to the capacity (1050b3–4).

Teleology has a pervasive role in this part of the chapter. I shall give three examples.

1. The reason why actuality is prior *in being* to capacity is that the presence and nature of the potentiality is teleologically explained by the goal in question (1050a9–19). We acquire (or possess) the relevant capacity in order to engage in the associated actuality. Introducing talk of teleology gives Aristotle a reason for taking actualities as prior in being (and not merely in account) to capacities. Teleology gives a real world basis for this claim. While more needs to be said about this, teleological concerns are central (it seems) for a proper understanding of claim [E].
2. Aristotle relies on teleological concerns when his comparison between processes and substances comes under pressure. After he has separated the activity of house building from the final product (the *entelecheia*: the house), he is faced with a problem. It appears that the goal is the house, not the process of house building. Aristotle, however, still wishes to maintain that house building (the activity, not the perfected state) is prior to (and more a goal than) the potentiality (1050a28). Can he consistently do so?

His explicit answer runs as follows: house building is more a goal than the potentiality because it occurs in what is being built and 'comes to be and is together with (*hama*) the house' (1050a29). But why should these factors be relevant to the goal-status of the activity of house building?[19]

[19] Aristotle's claim that 'house building comes to be and is together with (*hama*) the house' (1050a29) may mean no more than that the house building occurs in the same place as the house. For this use of 'together with' see *Ph.* 226b22ff., where, as Ross correctly notes, two things are said to be 'together' (*hama*) if there is one place which contains nothing but the two. So understood, the train of thought in 1050a28–9 is as follows: since house building occurs in what is being built and what is being built is in the same place as its successor, the finished house, the house building occurs in the same place as the house. If so, there is no suggestion (in this passage) that the builder's activity of house building continues (in some way or other) after the house has been completed. Sarah Broadie, in Chapter 9 of this volume, seeks to make sense of the latter idea as she (i) takes the type of togetherness to be temporal and (ii) understands all temporal togetherness as requiring extended co-temporality. However, even if (i) were correct, it is not clear that 'temporal togetherness' always requires extended co-temporality. For example, when in *MA* 702a19–21 a resulting action is said to occur 'together (*hama*), that is quickly' and to be 'together so to speak with the act of thinking', Aristotle appears to

Teleology provides a clue. In a teleological account (i) the goal of the house explains why we engage in house building (to achieve that goal) while (ii) we acquire the capacity to house build in order to build houses (to do the activity which leads to that goal: 1050a11–12). Further, the goal (the house) explains why the activity is the way it is (in order to achieve that goal) and the nature of the latter activity explains why the capacity to build houses is the way it is: it is what is needed if one is to build houses in the way we do.[20] If so, the teleological chain of explanation begins with the house, comes first to the activity of house building and only then moves to the capacity for house building. The relevant teleological order of explanation runs as follows:

House [A]
House building [B]
Capacity for house building [C]

The status of being a goal spreads from A to B down the chain of teleological explanation. B is more a goal than C (1050a27f.) because B is closer than C to A in the order of teleological explanation. The 'togetherness' of [A] and [B] is evidence for their relative positions in the order of teleological explanation and can be invoked to support the claim that, while A is the goal, B is more goal-like than C.[21]

Consider a case of goal-directed practical reasoning. Evans's final goal is, let us imagine, to climb Everest [A]. To achieve this he sets himself a number of intermediate goals: climbing K2 [C] and climbing the South Col of Everest [B]. Although Evans climbs the South Col in order to climb Everest, climbing the South Col is still a goal for him. Indeed, it is more of a goal for him than climbing K2 because (i) it is spatially closer to his final goal and (ii) he climbed K2 in order to achieve this as his next goal. In this case, as in Aristotle's example, [A] is the goal and that [B] is more goal-like than [C] in the order of teleological explanation.

be thinking (only) of immediate temporal contiguity (when nothing intervenes). Further, when in *Cat.* 14b24–6, two things are said to be together (*hama*) when their coming to be (*genesis*) is at the same time, there is no requirement that they continue to exist for all the same times. If so, the house and the house building might come into existence at the same time (the house then, no doubt, at some low level of actuality) even though the house (once completed) goes on existing after (see *para*, 1050a30) the process of building has finished.

[20] Reference to the intermediate goal of house building is important. Remove it and retain just the goal (the completed house) and one fails to explain why the capacity is as it is. For reference to the way in which houses are built is needed to explain the nature of the relevant capacity. Further, one acquires the capacity to build in order to build (1051a12f.) not in order to have a completed house. (One could acquire a completed house without becoming a builder!)

[21] Consider a non-teleological analogy: in geometry, the point may be the basic starting point (*archē*) but the line can still be more of a starting point than the triangle. For an example of this type, see *Ph.* 2.9.200a19ff.

3. In Θ.8 teleology underwrites Aristotle's central comparison between matter and form and the capacity for change and change itself. In both cases: (i) the end product (or result) is a goal; and (ii) since the end product is an actuality, the resulting actuality is a goal. The goal-like status of the actuality in the two cases allows Aristotle to present both types of result as actualities even though they differ ontologically: in one case there is an activity (e.g. displaying knowledge: 1050a18), in the other a state (being a house/ being a statue). Their shared goal-like status entitles him to speak of both as actualities. Thus, he notes:

> The result (in the case of action) is a goal, the actuality (*energeia*) is a result, and this is why the term '*energeia*' is predicated on the basis of what is the result (in the case of action) and extends to the perfected state (*entelecheia*).[22] (1050a21–3)

In this passage it seems that goal-like status of the results of actions and of perfected states entitles us to refer to both as '*energeiai*'. If so, teleological concerns ground the analogy between activities (the results of action) and perfected states.[23]

This point can be spelled out more fully. In the case of processes, the relevant capacity is defined as the capacity for a given type of process. What makes the process (type) the one it is is the goal (or result) that it achieves if all goes well (*Ph.* 202a25).[24] Since process-types are defined by their goal (or end point), the relevant capacity will be defined as the capacity to reach that end point. Here, the final cause (or goal) plays a role in defining both the nature of the process (to move towards that end point) and of the relevant capacity. In Aristotle's favoured example, house building is defined

[22] For a similar line of thought, see Θ.3.1047a30 where Aristotle notes that the name '*energeia*' is extended from a primary application to processes to refer (derivatively) to *entelecheiai* (that which has the goal in it: the perfected state). In 1050a21–3 he refines his earlier thought, noting that the term '*energeia*' is predicated of what is done in the case of action (either the action or the result of action) and together with this refers also to the perfected state. In this translation '*teinei eis*' can indicate 'refer to' (Pl. *Cra.* 439b10–c1) when words 'stretch to' objects. The prefix '*sun*' in '*sunteinei*' suggests (as in 1047a30) 'together with its basic reference'. The '*pros*' in '*pros tēn entelecheian*' may indicate the direction in which the term '*energeia*' is stretched when it refers to the perfected state. '*Teinei pros ti*', however, can mean 'come near to' or 'be like' (*Tht.* 169b; *R.* 584d). While I have followed Ross in using the first translation of '*sunteinei*,' an alternative would be to understand 1050a22–3 as follows: 'the term "*energeia*" comes to be like the term "*hē entelecheia*"'. Since the difference in sense between these two (acceptable) translations is not germane to my purposes, I shall not seek to adjudicate between them.

[23] The translation of '*energeia*' remains controversial. While I have used the term 'actuality' simply as a stand in for '*energeia*' in this essay, it seems important to capture the idea of something's being '*en . . . ergoi*' (literally 'being in or at work'). If so, the basic sense should (perhaps) be tied to activity but the term can extend to the product (work as in artwork).

[24] On this issue, see Charles 1984, 23ff.

as the process it is by its goal (a house) and the relevant capacity is defined as the potentiality to achieve this goal by building a house.

If substance is understood on a similar model, there should be a goal which reveals both its nature (conceived of as its perfected state: *entelecheia*) and the relevant potentiality (or capacity: its matter). Here too both actuality and potentiality will be defined (directly or indirectly) in terms of the relevant goal. The form will be the perfected state of the organism and the matter will be that which achieves that goal (if all goes well). The final cause plays a role in defining both the form and the relevant potentiality. The matter of a man, to use Aristotle's favoured example, will be defined as the potentiality to achieve its goal (e.g. the perfected state of being rational in certain ways) which is used to characterize the form in question.[25]

Teleology and Θ.6–7

In Θ.6 Aristotle lists five cases which fall under his favoured actuality/capacity (or potentiality) terminology (1048a36ff.):

(1) Building : what is capable of being built with
(2) Being awake : being asleep
(3) Seeing : what is capable of seeing
(4) What is separated from matter : matter
(5) What is worked up : that which it is not yet worked up.

The Greek term 'actuality' ('*energeia*') can, it seems, apply in some way to all the terms on the left of the colon and 'capacity' ('*dunamis*') to those on the right. But what is offered in Θ.7 and 8 (on the present account) is a philosophical explanation of why this is so: the terms on the left of the colon are (i) the goals of what is specified on the right and (ii) what enable us to define the relevant potentiality on the right. Indeed, this account offers a way of establishing the unity of Θ.6–8: the first chapter points to an analogous use of the terms 'actuality' and 'capacity' in a variety of contexts while Θ.7–8 provide an explanation of this based on the presence in (1)–(5) of a similar teleological structure.

Consider Aristotle's examples (3)–(5). The case of seeing is reintroduced in Θ.8 both as what explains why we have the relevant capacity (1050a10ff.) and as what defines that capacity (1049b20ff.). Similarly with the case of what is worked up from matter and the unworked matter from which it comes (1050a19ff., see 1049b27ff.). More generally, in Θ.8, matter is

[25] This formulation is intended to be neutral on the issue of the precise relative priority of goal and actuality. On this latter issue, see Gotthelf 1989; and Charles 1991, 108ff.

understood in terms of what will lead to the relevant product (the goal) if all goes well (1050a15ff.). This is taken to be true of cases involving external agency, such as the Hermes (1050a20) which is mentioned earlier in Θ.6 (1048a32) and internal agency (nature: 1050a19). In this way Aristotle can address the issues raised by examples (1), (4) and (5).

Case (1) is discussed as the first example in Θ.7 where the buildable is what is potentially a house. As I suggested above, what is capable of being built with (and the matter for a house more generally) is to be defined in terms of the art of building: it is the matter which is required for building to occur in the way which is intelligible given the goals of house building, and so on. Here, too, the term on the left of the colon is used to define the one on the right by noting the goal to be achieved by the builder. It is the latter which defines the art of building, the capacity to build and what is capable of being built with. But, that said, Aristotle struggles to show how building can be a goal even when the final goal is the house built (1050a23ff.). He needs this claim to sustain the comparative goal-like status of the term on the left over that on the right in this case also.

Aristotle does not return to case (2) in what follows, but elsewhere sleeping is taken as a case of having but not using certain capacities. If so, one can define sleeping as the state in which one is capable of achieving certain goals, the goals which define the activities in question. Further, as in the case of knowledge, Aristotle may intend us to think that being awake is the goal (1050a20–1) and that sleep is acquired (like the capacity to know) for the sake of this goal (see also 1050a9–14). He certainly sees sleep in this way elsewhere (*Somn. Vig.* 455b22ff.: 'the goal [of sleep] is being awake').

Only one example given in Θ.6 does not fit into this structure:

(6) The whole line: the half line when it has been cut away (1048a33).

This example does not re-emerge in Θ.7–8, perhaps because the teleological account developed there does not readily apply to it. It may be that Aristotle thought that if he had provided an account of matter in teleological terms, the case of the half line (the matter of the whole) could be understood by analogy. Or it may simply be that he focuses in the remainder of Θ.6–8 on the basic case of substance, where the formal and final goal are the same, leaving other types of example (where the explanation is not teleological) for discussion elsewhere.[26]

[26] In the case of the line, the line may be prior to the half line in the favoured mathematical order of explanation: see n. 21. Similar cases are discussed in (e.g.) *APo.* 2.11.94a25–35.

BEING AND BECOMING: THE ANALOGY BETWEEN PROCESSES AND SUBSTANCES CONTINUED

Being and becoming

If (as is argued above, pp. 171–3) Aristotle is addressing only question (B) in Θ.7, he will not be offering an account of the conditions necessary and sufficient for something's being potentially a B. More specifically, he will not be claiming in this passage that (for instance) it is a necessary and sufficient condition for something's being potentially a B that it retains the ability to become a B. If so, he can accept, as seems plausible, that in the case of (e.g.) a human, there can be stages in its development (e.g. when fully mature or aged) when it is still potentially a human but has lost its ability to become a human.

Indeed, Aristotle will not even be committed in Θ.7 to the weaker claim that the potentiality (or capacity) to be a B must include (at some point) the potentiality to become a B. For question (B) can be answered by pointing to a time at which A becomes potentially a B: by saying at what point A has the potential to become a B (of the type specified). While this may be the best way to mark out the point at which A is first potentially a B, it does not require that Aristotle understands the potentiality to become a B as part of the potentiality to be a B. For the latter potentiality may be more basic, not itself (even partly) defined in terms of the potentiality to become a B.

The latter possibility can be made more definite as follows: A may (first) be potentially healthy only when A is capable of being cured even though the latter capacity is not any part of what it is to be potentially healthy. So, for example, A's capacity to be cured may be a result of his more basic capacity to be healthy: his being in a state fit to receive health. What makes A capable of being cured is precisely his being potentially healthy. Indeed, this is a natural way to understand 1049a4–5:

> There is something which is capable of being cured and this is that which is potentially healthy.

On this view, the notion of being potentially a B is basic: the idea of A being in a state defined as fit for the relevant form of B In some cases, A is in this state and is actually a B. But in others, when A is potentially but not actually a B, A has the capacity to become a B But its having the latter capacity is a result of its having the potentiality to be a B. So understood, having the capacity to become a B (in the way specified) *is good evidence* that A is potentially B (in cases where A is not actually B). Indeed, one will acquire the ability to be a B first at the point at which one acquires the ability to become a B. But, even so, having the capacity to become a B need not be a constitutive part of what it is to be potentially a B.[27]

One will be attracted to the view just sketched if one thinks that the actualization of the capacity to become a B is *becoming* a B and not being a B, while the actualization of the capacity to be a B is *being* a B. For if capacities are defined in terms of their actualizations, these two capacities (or potentialities) will be different (even if, at certain stages in development, possession of one potentiality requires possession of the other). Thus, on one view, when A is potentially but not actually a B, A will have the potential to become a B.[28] (Acceptance of this latter point is not required for the present proposal.)

If the general view just introduced is correct, the potentiality to be a B can be present from the point at which A is capable of being turned into a B and persist through the period in which it is actually a B. So understood, there will be one account of the matter of B (as what is capable of being a B) which applies both to what is present before B comes to be and what is found when B actually exists. There is no need to see Aristotle as operating with two quite different notions of matter: the matter for becoming a B and the matter present when there is actually a B. Nor need one represent him as being committed in Θ to the project of accounting for the nature of either capacity (e.g. the capacity to be a B) in terms of the other (e.g. the capacity to become a B). If elsewhere he was attracted to the ambitious project of accounting for the latter capacity in terms of the former, this was the result of further theses he held, independent of his concerns in Θ.6–8.

The role of the analogy in the structure of Metaphysics *H and Θ*

If the account presented above is correct (in general terms), *Metaphysics* Θ will carry further the project of *Metaphysics* H.6 by showing how, in the case of substances, the relevant formal cause is identical with the final cause. For the actuality is the final cause and it is this which makes the potentiality the one it is. The potentiality is defined in terms of the goal it

[27] One way to make this possibility vivid would run as follows: one might take as the starting point of one's account the capacity to be a B and define the capacity to become a B as the *energeia* of the capacity to be a B at any stage at which the thing in question is not yet actually a B.

[28] For further discussion of this issue, see Charles 1984, 18ff.

will produce (if all goes well) and is what is required (teleologically) if that goal is to be achieved. Indeed, Aristotle, in introducing talk of actuality and potentiality, provided a way of conceptualizing matter and form so as to make his teleological commitments perspicuous and explicit (in the central case of substance).

On this account, when Aristotle remarked in Θ.6:

> All things are not said to exist actually in the same way, but only by analogy . . . for the relation is either that of process to potentiality or substance to some type of matter . . . (1048b8–9)

he was relying on an analogy between

the capacity for a process : the process itself

matter : substance

which he sought to underwrite through the teleologically based story told in Θ.7 and 8. Far from being the disappointing end to a failed attempt at explanation, this remark sets the target for the explanatory project undertaken in the next two chapters. When it is completed, the claim in H.6 (with which we began) will have been secured:

> What then is the cause of what is potentially F being actually F in the case of things that come to be over and above the efficient cause? For, nothing other is the cause of what is potentially a sphere being actually a sphere; rather this [i.e. the cause] is what it is to be for each of them singly. (1045a30–3)

For there will be one cause which makes the potentiality and actuality one: the formal cause (being an F) which is itself to be identified with the final cause. For this is (i) what the actuality is and (ii) what explains and defines the nature of the relevant potentiality.

In the course of Θ.7–8, Aristotle shows how the term '*ousia*' ('substance') in 1048b9 is to be understood. The relevant type of substance is identified with the form itself and not the composite (1050b2: 'the substance in the sense of the form'). Nor should this surprise us since in H.6 (1045a30ff.) talk of actuality (*energeia*) was taken as a replacement for talk of form and in Θ.8 (1050a16ff.) Aristotle regards the form as the relevant goal for the matter. In Θ.6 he is entitled to speak somewhat loosely of '*ousia*' ahead of the further clarification of this term given through his teleologically based account of its role in Θ.8. For there he wishes to argue for the priority of actuality over potentiality in a way consistent with his earlier comments on the priority of form (see, for example, *Metaph.* Z.10.1035a11ff.).

WHY REPRESENT MATTER AS POTENTIALITY AND FORM AS ACTUALITY?

There are several important consequences of presenting the matter of B as the potentiality to be a B. I shall mention four.

[A] *Continuity of matter liberalised (1)*: what is required for matter to persist in a new composite substance is just that no change or addition occurs which undermines the relevant potentiality to be a B. The matter of B (*to dunamei B*) will persist as long as the potentiality to be a B is preserved. The matter of B can change considerably throughout this period. It can persist without there being one quantity (one amount) of matter which persists throughout the generation and continued existence of the B in question, provided that no change, addition or subtraction undermines the relevant potentiality to be a B. However, that said, one does not need to go to the opposite extreme, thinking that, since the same matter does not have to remain unchanged in all respects throughout, it can be present in the new composite merely as something indefinite (or universal-like). This extreme, too, should be avoided. For if one is attracted to it, one cannot (easily) distinguish the case in which matter persists in the new compound from that in which a new entity comes into existence *de novo* with matter somewhat similar to the one it replaces.

[B] *Continuity of matter liberalised (2)*: what is required for the matter to persist in the case of a box is simply the persistence of the potentiality to be a box. In some cases, the actual wood may persist as what has this potentiality (e.g. as when placed as a door lintel). But in others, what is potentially a B may change in the process of formation and continuation of the composite (as in the case of the human mentioned above). At different stages in development, different types of matter can play the role of being potentially a B, provided that they all retain the relevant potentiality. Nothing is changed, added or taken away which removes that potentiality.[29]

This proposal allows the matter to be preserved even in cases where there is radical discontinuity. Thus, for example, in a thoroughgoing mixture, there need be no stable ingredients which pre-exist the mixture and persist throughout its existence. All that is required is that there is enough continuity in the matter for the same potentiality to persist. The mixture will be what is obtained when elements A and D mix (in a given ratio) to form a new matter capable of (1) sustaining some of the potentialities

[29] I am indebted to Dory Scaltsas for advice on this point.

of A and/or D (e.g. to heat/melt) and (2) returning to A and D on its dissolution. In such cases, one may not be able to state in general or informative terms what amount of continuity in the matter is required save in the tried and trusted formula: enough to sustain the relevant potentiality. If that potentiality is present both before and when the mixture is created, there will be the required continuity in the matter (what is potentially a B) without there being the same material stuff that persists throughout the process. Indeed, different types of material stuff can be at varying times the matter of a B. So understood, Aristotle's notion of the matter of a B is more abstract than might have originally been assumed.[30]

[C] *Essential unity of the composite confirmed:* thinking of matter as potentiality to be a B allows Aristotle to maintain the essential unity of the composite substance: it is what is found when the potentiality in question is actualized. Indeed, it seems essential to this matter's being the matter it is that it is what is actualized in this way in certain conditions. The matter cannot be properly defined except in terms of potentiality to be so enformed. This prevents the new composite from being a merely accidental unity in which matter (specified independently of its connection with form) happens at a time to play a given role: like Jones who is at some time given the rank of Regimental Sergeant Major.[31] Part of what it is to be this matter is to have the potential to be enformed in the relevant way: to be potentially a B (*to dunamei on*). The matter in question is intelligibly connected with the enformed composite in question (in virtue of the latter's connections with its form). The composite is not an accidental unity since matter and form (once conceptualized in this way) are essentially connected.

Indeed, this pattern will be reduplicated at all the stages in Aristotle's hierarchy where there is a combination of matter and form. Take the Θ.7 cases of box, wood and earth. Just as the matter of the box, is intelligibly connected with the form of the box, so too will the matter of the wood be intelligibly connected with its form – something will be the matter of wood provided that it has the potentiality to be wood (either when it is able to be made into wood by an actuality-inducing agent or when that potentiality is realized in actual wood). So too with the matter of earth

[30] I have argued for a similarly abstract view of prime matter in *On Generation and Corruption* in Charles 2004, 151–69.

[31] For a contrasting view, see the functionalist account of the realization of a mental description by a physical state.

if this too is made up by the imposition of a form on what is potentially earth. Each of these complex objects will be essential unities.[32]

[D] *Unity of definition sustained*: conceptualizing matter as potentiality and form as actuality allows Aristotle to underwrite the unity of definition of the composite in the following way: the actuality (given that it is the goal) explains why the matter (conceived as potentiality) is as it is by explaining both its presence and determining its nature (as both are required for that goal). In this way, the relevant goal both explains and defines the nature of the matter in the composite. If this is correct, the conceptual apparatus of actuality and potentiality (introduced in *Metaphysics* Θ) provides Aristotle with what he needs to provide an answer to questions such as 'What is man?' which at the same time explains why man's matter is the way it is. The formal cause is to be identified in this case with the final cause (as predicted in *Metaph.* Z.17.1041a32f., b4–9). The actuality is the one central feature (in this case) which makes man be the way he is (by teleologically explaining the presence and nature of his material being).[33]

INTELLIGIBILITY AND THE ROLE OF MATTER

Intelligibility and Aristotle's metaphysical proposals

The four points in the last section have a common source: the matter of a composite is what it is because it is teleologically required for the goal expressed in the actuality. This understanding of the matter allows for considerable flexibility in what constitutes the matter of the composite (see [A] and [B]) while making the unity of the composite fully intelligible in the ways captured in [C] and [D]. From this perspective the nature and unity of the composite becomes intelligible and perspicuous. One can see why its matter is as it is and why, given that its matter is understood in this way, the compound of matter and form is non-accidental (and captured in a properly unified definition). The nature of the matter and the composite are clear to us once we follow Aristotle's advice and see them as teleologically explained and defined by the presence of the relevant goal-laden actuality. In its light we can see why matter and composite are as they are (when all goes well).

[32] Fire, at the bottom level, will not be a unity of this type since it does not require an actuality-inducing agent to be what it is. The issue of what type of phenomenon fire is lies outside the scope of this study.

[33] So understood, the actuality is the basis for a definition of man given in a style which stems from Aristotle's views of definition in the *Posterior Analytics*. For more details, see Charles 2000, chapters 8–11.

The type of intelligibility which Aristotle seeks is distinctive. It cannot be secured simply by setting out an efficient causal story of what happens (in this case), still less by specifying the materials required for the composite in teleology-free terms. For both these routes would fail to explain why the organism has to be made up or brought about in the way it is. To meet this latter requirement Aristotle introduces the idea of a goal which simultaneously defines the nature of the organism and explains why it is the way it is. As in *Posterior Analytics* 2.2 (90a14–15) there is one feature which answers both the definitional (what is it?) and the explanatory question (why is it the way it is?). If one grasps this, the nature of the matter and the composite will be properly intelligible.

Aristotle's master-builder (or master mason: see *Metaph.* A.1.981a30ff.) exemplifies the kind of intelligibility sought. He (or she) knows why the house has to be built the way it is (given what the house is designed to do) and why it has to made up from some materials rather than others (if these goals are to be met). He (or she) also understands why in the finished house these materials have to be organized in the way they are. His (or her) type of understanding will not be available to one who knows only what steps have to be followed to construct a house or that a house will exist when certain pieces of wood and stone are organized in a given shape (characterized in a non-teleological fashion). For that person will not know why the house has to be organized in the way it is or why certain steps are required to build it. The master-builder's perspicuous understanding of these issues, by contrast, indicates the type of grasp the successful metaphysician will have of his subject matter.

Intelligibility and matter

It was acknowledged above (pp. 193–4) that it will not (in Aristotle's account) be possible to state what the relevant matter is (in the fully intelligible and general way required) without reference to the form in question. To this we might add a further point: it will not be possible to state intelligible and general principles of construction required to obtain the composite without reference to the form in question: they will be the ones required for this form to be present. These claims follow from the demand that it be intelligible why the relevant matter is present and why it is changed or altered in the way it is for its potential to be realized. One cannot state in a sufficiently general or perspicuous way why this matter is required or why it is acted on in the way it is without reference to the form in question. And, for these reasons, one cannot state in sufficiently general or perspicuous terms what the matter in question is.[34]

The point just made is, however, fully consistent with the truth of a further claim: one can state (in particular cases or generally but in a less than fully perspicuous way) what changes will undermine the relevant potential and which will sustain it or allow it to develop. The latter changes are ones which will, for example, in particular cases, change the matter so that it degrades and loses its potential to be enformed. Equally, one could (in some particular case) begin with the capacity to be a house and investigate what features of the bricks are present when this capacity is present and what alterations would remove it (e.g. being melted). But such specifications of the matter would fail either to specify what types of matter in general have to be present for there to be a house (to unify the class) or to explain why it is just these types of matter that are needed. For one would lack the answer to these questions provided by reference to the teleological cause: it is the type of matter required for the goal in question. Remove that and one fails to make it fully intelligible why it is these types of matter (or changes in matter) that are present in a house. The limited and less than fully perspicuous explanations just mentioned will not concern the matter of a B (since that is conceptually connected with the potential to be a B) but rather the ingredients of that matter: the earth and water that compose blood (*Mete.* 389a19–22) rather than the blood thus composed (*PA* 2.4.650b14–21). For the matter (e.g. the blood) is what it is in virtue of its connection with the form in question.

The structure just outlined can apply even to the more radical case of mixture. Here, too, one may not be able to specify in general or perspicuous terms what the principles of mixing are except in terms such as those required for the relevant form. Nor need one be able to state in these terms what the ingredients of the new mixture are (let alone its *logos*) without reference to its form: the ones that are required for that form to be actualized. These points are, of course, important. But, as they stand, they are consistent with its being possible to state (in particular cases or in a less than fully perspicuous way) what changes will undermine the relevant potential or what is required to sustain it in terms which do not advert to its potentiality for form. Such specifications (in a non-form involving

[34] There is, nonetheless, more to the specification of the matter in question than the bare idea of its potentiality to be enformed. For Aristotle introduces the idea of matter as that in which no further changes or additions are required if the capacity is to be present (and the strictly correlative idea of there being changes or additions which will undermine that capacity and destroy the matter in question: 1049b10–11).

chemistry of mixture) would fail to make it intelligible to us why just such ingredients are required if the mixture is to exist. But nothing so far said rules this out. If it is to be ruled out, Aristotle would need to show that no chemistry of this type of mixture could be developed. It is doubtful whether he attempted to do that. But, more significantly, from the present perspective there was no need for him to do so. For such a chemistry would fail to give the kind of understanding of the phenomena which is only available from the downward perspective.

On this understanding, there need be no conflict between the form-involving 'top-down' approach to matter, which makes it fully perspicuous why the matter is as it is (in the ways indicated on pp. 193–4) and a 'bottom-up' account which begins with matter non-teleologically characterized and works in the direction of the form. For even if the latter account could be completed, it would not yield the fully perspicuous understanding we seek of the matter of the composite. At best, it would yield a set of conditions sufficient for the presence of blood. It could not tell us what blood is, why it needs the constituents it has or why they have to be organized as they are. The 'bottom-up' account would fail to specify the essential features of the matter of blood: those required for the relevant goal. It could only describe in non-essential terms what the matter in question is like: hot, wet, and so on.[35]

INTERIM CONCLUSIONS

In this chapter, I have tried to understand why Aristotle sought to conceptualize matter and form in terms of actuality and capacity (or potentiality) in parts of Θ.6–8. My suggestion is that he did so to capture the importance of teleology for a proper understanding of the required ontology. While I have not attempted to discuss the details of Aristotle's teleology, it provides the key to a proper understanding of these chapters of his *Metaphysics*. Or so I have argued.

[35] While it would take (at least) a further paper to assess how far Aristotle actually pursued the proposal just sketched, my aim here is a modest one: to show how the view of matter that is developed in *Metaphysics* Θ.6–8 is at least consistent with the possibility of such an upward 'material' story. While I began to sketch this type of account in Charles 1988, it needs to be developed further and in more detail.

CHAPTER 9

Where is the activity?

(An Aristotelian worry about the telic status of *energeia*)

Sarah Broadie

The question of my title comes to the surface in *Metaphysics* Θ.8.1050a4–b4. Aristotle is seeking to establish that actuality or activity (*energeia*) is prior in substance (*ousia*) to potentiality (*dunamis*). More precisely, he is trying to show that a given *energeia* is prior in *ousia* to the corresponding *dunamis*. In reaching his conclusion Aristotle relies on the curious assumption that in the case of a transitive activity, for example the activity of a builder transforming materials into a house, the activity itself is located not, as we might expect, in the agent, but in the patient – in this case, in the materials which are coming to be the house (1050a30–4). My object here is to examine this premise and its contribution to the surrounding argument. It will be necessary, however, to look at the whole passage, and in doing so one will find it natural to touch on a few difficulties which do not immediately bear on the question of the location of an activity. Along the way I shall also try to bring out some of the philosophical issues involved.

The demonstration that *energeia* is prior in *ousia* to the corresponding *dunamis* is third in a battery of arguments showing that *energeia* is prior to *dunamis* in many ways: in definition, in time (although this conclusion is qualified) and in *ousia* (1049a4ff.). Now, it is easy to show that *energeia* is prior in definition, for obviously a complex expression such as 'the *dunamis* of *F*' cannot be clarified unless one first clarifies the embedded expression '*F*', which in this context stands for the *energeia*, whereas the clarification of '*F*' does not presuppose clarification of the complex expression. By contrast, it is not so easy to prove the other two kinds of priority. Since the *dunamis* of a thing and its *energeia* are both temporal entities – both exist in time (*en chronōi*) – it makes sense to ask which is temporally prior, and,

I am delighted to have an opportunity to contribute to this volume in honour of Allan Gotthelf, who has done so much to promote the philosophical study of Aristotle's biology.

as Aristotle's argument shows, there are considerations favouring each of the rival answers. Rather in the same way, *dunamis* and *energeia* can each be viewed as the substance (*ousia*) of a thing, at least if we attend to the close connection Aristotle draws in many places between *dunamis* and matter on the one hand, and *energeia* and form on the other. For he indicates, again in many places, that the matter of a thing and its form each have a claim to be considered its substance. It is precisely this that prompts the question 'Which has the stronger claim?' both for the matter/form pair and for the *dunamis*/*energeia* pair which concerns us here.

Plato, after all, had suggested that *ousia* is *dunamis* to act or be acted upon.[1] And although Plato himself made this suggestion without at the same time considering any distinction between *dunamis* and *energeia*, it is plausible that, once that distinction was made, some reasons would be found for holding that, of the two, *dunamis* is the more substantial. After all, in many rather obvious cases the *dunamis* remains when the *energeia* has come and gone: for example, animals have the *dunamis* of seeing regardless of whether they are actually seeing. Is not this superior permanence a sign of superior substantiality? Similarly, there are obvious cases in which the *energeia* in fact just is a process of change or movement or becoming, whereas the same cannot be said of the corresponding *dunamis*. From a broadly Platonic point of view, this special association of *energeia* with change surely suggests that *energeia* is less substantial. This, I think, is the philosophical background to be borne in mind as we turn to the detail of Aristotle's argument in *Metaphysics* Θ.8.1050a4–b4.

Here, Aristotle reasons as if he can secure his conclusion that *energeia* is prior in *ousia* to *dunamis*, simply by showing (1) that *energeia* is prior in form (*eidos*), and (2) that *energeia* is the principle (*archē*) of *dunamis* in the sense of being its end (*telos*). The same initial examples – the adult male (*anēr*) and the child, the human being (*anthrōpos*) and the semen – serve to illustrate the distinctions (1050a5–6). The two questions, one concerning priority in form, the other concerning the role of *telos*, are closely linked in the argument of our passage, but they nonetheless present distinct standards for deciding which candidate is prior to the other in *ousia*. In proposing the *energeia* as *telos*, Aristotle is in fact proposing that the *energeia* is an origin or principle (*archē*) in relation to the *dunamis*; thus the *energeia* is metaphysically superior and metaphysically more powerful and fundamental. Of course, in order to obtain his desired result, Aristotle must interpret *archē* in terms of final causality, and he gives no grounds

[1] *Sophist* 247d8–e4.

for preferring this interpretation. We may see this as a weakness in his argument, but if Aristotle is mainly addressing Platonists he can surely rely on them to grant him the point. Platonists, of all people, should be ready to identify *principle* with *final cause*. If, instead, one were to understand *archē* here as efficient cause (that *from* which a change arises) and were then to apply the same reasoning, one would conclude, on this ground at least, that *dunamis* is prior to *energeia* in respect of *ousia*. The result would be a clash between the two standards for priority in *ousia*. For the doctrine that the *energeia* is prior to the *dunamis* in form (*eidos*), cannot be pushed aside.

I suppose 'form' here means the version, aspect or mode of a thing that one has to attend to and grasp if one is to be in the best cognitive state regarding such a thing. Most kinds of thing occur in two versions, potential and actual. These seeds in this jar are rightly called and classified as 'marigolds' when put side by side with the seeds in this other jar, which are rightly labelled 'poppies'; this sleeping person is a violinist, while this other sleeper is a trombonist; this instrument in the kitchen drawer is an egg-whisk, while its neighbour is a mandolin for slicing onions; these dry little sachets are tea, while these other ones are coffee. Yet if you want to *know what* a violinist is and tell the difference from a trombonist, you must wake the sleeping musicians and hear and see them actually play, and so on with the other examples. Neither perception nor intellect can grasp what is distinctive about marigolds and what is distinctive about poppies as long as no seeds of the two kinds grow up into flowering plants. Thus knowing what a thing is – grasping its *eidos* – inextricably depends on accessing the thing in its actual or active mode. So if we were to interpret the notion of *archē* in terms of efficient, rather than final, causation, then – at least in the case of the plants and the many other cases like them – we would be identifying the *archē* with potentiality, and not with the mode in which the thing in question is immediately and intrinsically knowable. At the same time, we would be treating its actual and immediately knowable mode as a derivative phenomenon, since something else would be *archē* of this and this would be *archē* of nothing. In short, the *eidos* or knowable aspect of a thing and its *archē*-aspect would come apart from each other. In this situation we would have to abandon the assumption that each of these in its own way spells priority in *ousia*. For otherwise we obtain contradictory results: by one standard, *energeia* is prior in *ousia* to *dunamis*; by the other, it is posterior. If forced to choose, we might, I think, retain the coupling of *ousia* with *archē* and reject the coupling with *eidos*. We might equate substance with power:

power unknowable in itself, and identifiable at best as the origin of certain *in*substantial effects.[2]

It is understandable, then, that Aristotle interprets *archē* as the *telos*, thereby ensuring that his two standards for priority in *ousia* in Θ.8 yield consistent results. We cannot, however, pass on without briefly noting a fact which has puzzled several commentators. In Δ.11.1019a1–14 Aristotle formulated priority in respect of *ousia* (*kata phusin kai ousian*, 2–3) as follows: X is prior in *ousia* to Y if and only if X can be without Y, while Y cannot be without X. But this principle seems to be absent from the part of Θ.8 which concerns us. Or is it somehow implicitly present in the two standards which we have been considering? But how can the Δ principle yield an acceptable result for this part of Θ? For on the face of it, it is false that the actuality of a thing can be without its potentiality, and that the potentiality cannot be without the actuality. Here, I shall not try to take up a position on this question of reconciling priority in *ousia* according to Δ.11 with priority in *ousia* according to our portion of Θ.8.[3] Instead, I shall continue to examine the latter just in its own terms.

Returning then to our text, we see Aristotle making his point with different kinds of examples of *dunamis* and *energeia*. Becoming (*genesis*) gives us one kind, exemplified by the pairs child/adult male, sperm/human being (1050a5–6). Then we have capacities and their corresponding activities, with illustrations drawn from natural capacities such as the faculty of vision, and from acquired ones such as expertise in building and expertise in theorising (10–12). Between them these examples also display the contrast between intransitive activity, such as seeing and theoretical thinking, and transitive activity such as house building, where an agent works on a patient to produce a distinct object or effect. Aristotle then emphasizes that his point also holds for a perhaps less obvious kind of transitive activity: the kind that results not in a static hylomorphic compound such as a house, but in an activity. For example, teachers feel that they have delivered the end for

[2] In response to the interpretation of *eidos* which I adopt, one may wonder in what way priority in *ousia* in terms of *eidos* differs from priority in definition (*logos*), which figures as a distinct type of priority at 1049b10–11. The question was raised by Anna Marmadoro in comments on an earlier version of this chapter. In reply: one can follow the route which Marmadoro herself took in her comments, which was to interpret *eidos* in our passage in a more purely metaphysical fashion: this is by contrast with the epistemic interpretation which I employ here. Alternatively, one can emphasize the difference between the analytic or formal (in the modern sense) observation that the complex expression '*dunasthai* ϕ' is logically posterior to the simple 'ϕ' for any value of 'ϕ' (the grasp of which does not depend on knowing the meaning of '*dunasthai*'), and the logically less trivial observation that in order to know the differences between things one must attend to cases that are actual rather than potential.

[3] On this question, see Makin, 2003, 209–38, and 2006, 192–7.

which they were hired when they show their students actually performing (17–19). Presumably the students' skilful performances may be intransitive ones such as singing or solving a theoretical problem, which result in nothing further, or they may themselves be transitive ones such as converting clay into a pot or bricks into a house.[4] While the teacher's achievement is rational and technical, we are now told that the point brought out by this example applies also to nature (19). (Thus the *telos* of the parent is not just the offspring considered as a hylomorphic compound, but the offspring's natural activities.)

Aristotle then tries to make things clearer with an illustration, but both the illustration and his use of it remain obscure to us. He says that we must think of the teacher's *telos* as consisting in the student's actual performance, because otherwise it will be like the Hermes of Pauson: just as people could not tell whether the Hermes was 'inside or outside' so they will not be able to tell whether the knowledge is inside or outside (19–21). It seems that Pauson painted *trompe l'oeil* effects,[5] and perhaps his Hermes was painted so that the figure seemed to be both here and there, rather as figures by Escher appear as both ascending and descending the same staircase. But this is quite uncertain. As for the predicament of not being able to tell whether the knowledge is inside or outside: are 'inside' and 'outside' said in reference to the pupil or to the teacher? Either way, there appears to be a 'location problem'. This is interesting as an overture to the next main stage of the argument, where Aristotle will announce that a transitive activity has its location in the patient. As regards the location of knowledge in the teacher-case, I am inclined to think Aristotle means this: rather than go to and fro between saying that the teacher's knowledge is inside the teacher – since he has mastered the subject – and saying that it is outside the teacher – since it is in the pupil – we can settle the question rationally once we see that what counts as the teacher's knowledge *par excellence* is not his mastery of the subject or his capacity to teach, but his achieved goal or *telos* as teacher; for this is clearly the pupil's activity, and it is unequivocally outside the teacher.

The above may seem too far fetched. Admittedly, only an Aristotelian philosopher could think of equating *the teacher's knowledge* in the strictest sense with (as we can put it) *what the teacher's knowledge is 'all about'*, i.e.

4 Aristotle calls such a resultant activity '*kinēsis*' at 1050a17, but there seems to be no reason to think that here he restricts the type to the strictly conceived *kinēseis* of Θ.6.1048b18–35, which (because they are contrasted with *energeiai* strictly conceived) cannot themselves be classified as *energeiai*. On the place of 1048b18–35 in the *Metaphysics*, see Burnyeat 2008.

5 See the comment of Ross 1924 on these lines.

the pupil's educated performance. But this is only one step more strange, by the standards of ordinary discourse, than equating the scientist's knowledge in the strictest sense with the scientist's activity in using that knowledge in actual demonstration; and we know that Aristotle maintains this latter equation. (Here I am thinking of the scientist as not, as such, a teacher of science; thus the scientist's knowledge in the strictest sense is considered as terminating in the scientist.) And anyway, would it really be so strange if a master cake-maker, surveying a board on which were displayed examples of every cake in his or her repertoire, said: 'All my knowledge, all my art, all my skill is laid out on that table!'?

Fortunately, uncertainty about these details does not obstruct our view of Aristotle's main thought, which is that in cases like that of the teacher the *telos* is an *energeia* – the *energeia* of the pupil (or the pupil-analogue). He supports the point with two premises of a syllogism: (1) the *ergon*, or typifying result, is the *telos*; and (2) the *energeia* is the *ergon*. From this it follows that the *energeia* is the *telos*. The first premise is taken to be self-evident; the second is bolstered by the etymological derivation of '*energeia*' from '*ergon*'. And then, in the very same sentence (21–3), Aristotle tries to reinforce the tie between *ergon* and *telos* on the one hand, and *energeia* on the other, by observing that the *energeia sunteinei pros tēn entelecheian* ('the activity strains towards fulfilment'). Here he is invoking yet another etymological connection, that between '*telos*' and '*entelecheia*'.[6]

Before continuing, I must explain how I punctuate the sentence '*to gar ergon telos . . . pros tēn entelecheian*' (21–3). Contrary to most interpreters, I understand the clause beginning '*diho*' in line 22 as a parenthesis ending with '*to ergon*' at 23. (Most interpreters take '*diho*' as governing the entire rest of the sentence.) Thus I place dashes or the like after '*ergon*' in 22 and before *kai* in 23. The subject of '*sunteinei*' is therefore '*hē energeia*' at 21–2. In other words, the phrase '*sunteinei pros tēn entelecheian*' is predicated of activity itself, not of the expression 'activity' (*t'ounoma energeia*, 22), as most interpreters have it. The remark is metaphysical, even though it also exhibits an etymological connection.

Aristotle's thought is particularly dense and complicated in this sentence. In the first clause (for which evidence is adduced in the parenthetical second one), he equates the *energeia* with the *ergon*, and so, by implication, with the *telos*. Here the word '*energeia*' surely refers to an activity such as that of the performing pupils, which is the ultimate end of the teacher. (It

[6] See Ross's note on 1047a30.

may also refer to the actuality of an object such as a house considered as the *telos* of the builder.) Certainly, this occurrence of '*energeia*' ought not to refer to the teacher's activity of teaching (or the builder's of building), because these types of activity tell *against* the very doctrine being forged in this part of the sentence – the doctrine that *energeia* is the *telos*. Transitive activities such as building and teaching are not ends in themselves. However, as subject of the next clause, in the context of the statement that the *energeia* strains towards fulfilment, the word '*energeia*' refers to what is surely *not* the fulfilment or ultimate *telos*, since it is a straining towards this. And now the activities of teacher and builder are perfect examples.[7]

Here, then, is Aristotle's problem. He is aiming to establish that *energeia* is prior in *ousia* to *dunamis*, by way of the intermediate theses that *energeia* is prior in respect of form, and that *energeia* is the *telos* of *dunamis*. Several kinds of example support his desired conclusion, but, on the face of it, one particularly prominent and obvious kind of example points the opposite way. This is the kind where the *energeia* is a transitive activity: the activity of an agent on a patient for the sake of some ulterior result.[8] Since the activity in such a case is clearly subordinate to the ulterior result, why should we view the activity as a *telos* any more than the agent's mere *dunamis* for that activity? Why should we agree that this activity is a *telos* of the *dunamis*? Why should we not demote the activity to the same status as the *dunamis*, so that both are merely subordinate to the one and only real *telos*, which is the ulterior result? But if we go down this path with the transitive activities, how can we still be justified in maintaining that activity in general is prior in *ousia* to *dunamis*?

These, it seems to me, are the questions in play at that point in Θ.8 where Aristotle digs in his heels and insists that, in the case of transitive activities, the activity is the *telos* more than the agent's *dunamis* is (1050a27–8; see section V in the appended Analysis).[9] Here is what he says:

[7] Thus '*hē energeia*' at 21–2 refers both to the end of process and to the process itself. This manoeuvre is facilitated by the assimilation of *energeia* to *ergon*, since the latter term can signify either the product or the productive action. In this way Aristotle links *energeia* equally to the two contrasted aspects: process towards an end, and an end achieved. The paradox of this double reference is (I think) what has led scholars (e.g. Ross, Jaeger, Tricot in his translation, and most recently Makin (2006)) to make '*t'ounoma energeia*' the subject of '*sunteinei*', and to understand '*sunteinei pros*' + accusative as, for example, 'tends to signify'. This interpretation removes the paradox, but it rests on a dubious sense of '*sunteinein*'.

[8] The *dunameis* corresponding to *energeiai* of this type are the subject of the first two chapters of Θ.

[9] Although Aristotle will soon touch on the *patient* of a transitive activity, and hence by implication on the patient's *dunamis* to be acted upon, the present argument plainly focuses on the *agent's dunamis*.

But since[10] in some cases (*tōn men*) the use is the last thing (e.g. seeing in the case of the sense of sight; i.e. nothing else beyond (*para*) this [sc. seeing] comes about from the sense of sight as its product (*ergon*), whereas in some cases (*eniōn de*) something does come about (e.g. from the house-building art there comes about a house beyond the house building) – all the same, this does not alter the fact that <the activity> is, in the former case the end (*telos*), and in the latter case more of an end than the potentiality is. (1050a23–8, my trans.)

Aristotle here makes two assertions, each contradicting a mistaken view which he thinks we might find tempting once we focus on transitive activities which result in *erga* beyond themselves. First he declares that even though there are those transitive activities, this does not alter the fact that (*homōs outhen hētton*, 23) in the case of *in*transitive ones, the activity itself *is* the *telos*.[11] And secondly he declares that, in the transitive case, the activity is more the *telos* than the *dunamis* is. Now it is taken for granted that the ulterior *ergon* of a transitive activity is the *telos* of the whole affair. What Aristotle fears is that we shall become so fixated on transitive activity, and particularly on its relation to the ulterior *ergon*, that we shall end up believing that an ulterior *ergon* (or something similar) is the only kind of *telos*. In that case, we shall be driven in the direction of denying that the *in*transitive activities are *telē*, since in their case there is no ulterior *ergon*.

True, only a short time ago it seemed self-evident that the intransitive activities are *telē* in relation to possession of the corresponding *dunameis*: it was simply stated, as needing no argument, that we have the capacity of vision in order to see, and so on, not vice versa (1050a10–13). But presumably there was no suggestion at that point that anyone might think that *some* activities, namely the transitive ones, are completely devoid of telic status. Such a thought arises only when we make transitive activity our main model, and we focus on the obvious fact that this activity is subordinate to the ulterior *ergon*. By a logical extension, we now cease to be so sure about the telic status of even the intransitive activities. This, it seems to me, is the movement of thought that Aristotle wants above all to block, and he attempts, it seems to me, to block it by insisting that, even in the transitive case, 'end-hood' (or finality) is not, after all, the absolute and exclusive prerogative of the ulterior *ergon*: the transitive activity itself is a sort of end; it is more an end than the *dunamis* is. If we accept this, then we are endowing activity as such with some degree of telic status, and

10 *Epei* here does not introduce a premise from which the apodosis is supposed to follow: it signals that the clause it introduces motivates what is said in the apodosis.

11 It is not clear whether he means that it is the *telos* of the corresponding *dunamis*, or whether it is a *telos* in some more absolute sense.

in that case we have reason to continue to affirm that intransitive activities are ends.

I shall now look at the reasoning by which he supports his doctrine that the transitive activity is more a *telos* than its *dunamis* is. Then (and this will be the conclusion of the paper) I shall indicate what is at stake in Aristotle's reaffirmation that the intransitive activities are really and truly ends.

Aristotle gives two considerations in favour of viewing transitive activity such as house building as more a *telos* than the *dunamis*: first, this activity is in the patient, in what is being worked upon by the agent so as to produce the house; and secondly, the activity 'comes into being, and is, simultaneously with the house' (28–9). Now, the doctrine that locates transitive activity in the patient, not in the agent, was forged in *Physics* 3.3, where it was shown that the agent's actual agency and the patient's actual patiency are one and the same activity viewed from opposite perspectives; that this activity is a *kinēsis*, a process of change or movement; and that the *kinēsis* is *by* and *of* (but not *in*) the agent, and *in* and *of* (but not *by*) the patient (202a14–b22). 'In' and 'by' are understood in such a way that the statement that a *kinēsis* is *in* X implies that X moves (English intransitive verb = *kineitai*, *se mouve*, *si muove*, *movetur*), whereas the statement that a *kinēsis* is *by* X carries no such implication. In this way Aristotle safeguarded his doctrine that some movers are unmoved.[12] But *Metaphysics* Θ.8 shows no sign of anxiety on that score. Instead, what motivates Aristotle here is the thought that locating the activity of house building in the patient not only separates it from the agent in whom exist the skill and *dunamis* of house building,[13] but also – which is more important – effects a rapprochement between the activity and the resulting house. By contrast, in the run up to the *Physics* 3.3 passage – in *Physics* 3.1 where he establishes his definition of *kinesis* – he draws a strong distinction between house building and house (201b11; see also 3.2.201b31 and 202a1, where the actuality of the process by comparison with that of the result is said to be only *energeia tis*, 'actuality – sort of').

This brings us to the second thing he says in the hope of making it plausible that the transitive activity is more a *telos* than the *dunamis* is: 'it [sc. the activity of house building] comes into being, and is, simultaneously with the house'. At first this seems ludicrous – he is clutching at a hopeless straw. For one thing, the activity of house building cannot come into being. Insofar as it is identical (see *Ph.* 3.3) with the process in the materials by which they become a house, the activity of house building *is* a coming into

[12] Cf. *Ph.* 3.201a19–27; 202a3–6, 30–1; see also *de An.* 3.2.426a2–6. [13] Cf. Θ.1.1046a19–27.

being – of the house; and Aristotle has shown in *Physics* 5.2.225b10–23, that there cannot be a coming into being of a coming into being. But then what happens when the builder starts his work after resting? First he is inactive with respect to house building; then the activity starts. Aristotle, however (if we go by what he says elsewhere), would not classify *starting to build* as the coming-to-be of the building activity: not every transition from the non-being to the being of something counts as a genuine Aristotelian coming-to-be.[14] But even if he were to waive his rule here, allowing that house building 'comes-to-be' when the agent passes from potency into act, this would not help. For the builder's commencing to build is not simultaneous with the final birth of the finished house, nor with the process by which the house gradually comes to be.[15]

It might seem charitable to interpret Aristotle as saying in a rather loose way: 'the activity of house building is a coming-to-be which develops or unfolds simultaneously with that process which is the coming-to-be of the house'. As well as making sense, this immediately follows from the Aristotelian doctrine I have just mentioned: that in transitive action the activity of the agent is identical with that of the patient. But the charity falls short. For what Aristotle's argument needs is a step that helps to make it plausible that the transitive activity is the *telos* more than its *dunamis* is. From that point of view, merely reflecting that ongoing transitive activity is simultaneous with the coming-to-be of the *ergon* gets us no further forward.

We must somehow make sense of Aristotle's statement that the activity of house building is or exists – not merely, as we all know simultaneously with the *genesis* of the house, but – *simultaneously with the house itself: with the existent house once it is actual.* But how could he state this when only six lines earlier he spoke of the activity as '*straining towards* fulfilment'?

[14] Cf. *de An.* 2.5 (although the argument there concerns *alloiōsis* rather than *genesis*). See also *Metaph.* B.5.1002a30–4; H.2.1026b22–4; H.3.1027a29–30.

[15] So far as I can see, the close discussion of Makin 2006, 200–4 does not explain the simultaneity claim. David Charles, on pp. 184–5 n. 19 of his contribution to this volume, takes it to mean no more than that nothing comes between the process of house building and the existence of the house. (As he says, 'on this interpretation, there is no need to suggest that the builder's activity of house building continues (in some way or other) after the house has been completed'. Precisely such a suggestion is what I try to make plausible in the main text below.) Charles bases his interpretation of *hama* in 29 on *Ph.* 226b22ff. However, this tells us only how Aristotle understands the spatial *hama*, not the temporal, and the latter sense (as Charles by his choice of words allows) is the one in play in our passage. Temporal *hama* is understood in terms of strict simultaneity (i.e. not mere temporal contiguity) at *Cat.* 13.14b24–6 and *Ph.* 4.10.218a25–6. At *MA* 8.702b19–21, to which Charles also refers, the context in fact suggests that the *hama* (here, clearly temporal) implies something like strict simultaneity between the thought/perception and the movement, since the former guides or controls the latter, which it could not do if the two were merely temporally contiguous; cf. 702a15–16.

This straining cannot coexist with the actual house. Building activity is present only because and when the house is not yet present, and the house is present only because and when there is nothing more for the builder to do. This is true so far as it concerns the *ongoing* house building. But surely we can suppose Aristotle to be talking about house building as *having been carried out*. Why should he not think of the transitive activity as having two faces: the *ongoing* and the *having been completed*? Of course *having been completed* does not inevitably ensue: there can be interferences. But *having been completed*, when it does ensue, ensues precisely where and when it belongs. Nothing could be less adventitious. Consequently, it seems blind or arbitrary to reason as if an activity no longer ongoing is an activity in no way present – as if *ongoing* were the only aspect of transitive activity.

A particular process of house building in its *having been completed* mode is simultaneous with a certain house in just the way the form in a hylomorphic composite is simultaneous with the composite. A form of this kind is real when and only when realized in matter, and precisely then there exists the matter–form composite. In fact, without more ado we should surely equate the *having been completed* transitive activity with the form of the hylomorphic *ergon*. For the *having been completed* activity of building is what makes the difference between the matter's not being a house and its being one. Thus the statement that the house is a *telos* beyond (*para*, 1050a25–6; 30; 35) the activity of house building turns out to be true only with qualification. The actual house is beyond the ongoing activity, but it intimately includes the having been completed activity.

In this way we can make sense of Aristotle's assertion that the *energeia* of house building is a *telos* more than the corresponding *dunamis* is. That *dunamis* exists for the sake of a house or houses, which figure here as the uncontroversial *telos*. Equally uncontroversially, any house is other than and external to the *dunamis* for house building. Thus this *dunamis* is unequivocally subordinate to something *beyond* itself. By contrast, the ongoing *energeia* of house building occurs for the sake of a house, but equally it occurs so that its own having been completed version should ensue. In a way, then, this *energeia* occurs for the sake of *itself*. True, it occurs for the sake of the house, a thing made of bricks and tiles; and, even in its having been completed version, the *energeia* of house building cannot itself be the kind of thing that is made of bricks and tiles. But this is because in its having been completed version it is the form of an actual house. So the *energeia* of house building is for the sake of something conceptually distinguishable from itself – an actual house, a composite substance. Even so, the completed *energeia* will be *where* the actual house will be. For of course the composite substance cannot be beyond its own form. In a way, then, the concrete house, which is the ultimate end of house building, is not beyond the activity itself, even though from the point of view of the *ongoing* activity the end is plainly quite other and beyond. At any rate, these lucubrations should be enough to persuade us to grant transitive activity a telic status higher than that of the corresponding *dunamis*.

It must all the same be admitted that the appearance is strong that transitive activity is no more of a *telos* than the mere *dunamis* for it. For the *ongoing* aspect is the obvious one, and it is essentially interruptible – which is to say incomplete (*atelēs*) – while it occurs.[16] If the activity is interrupted and the *ergon* beyond it fails to be produced, the activity has achieved little more than the inactive potentiality. Suppose now that one were to yield to this unmitigated appearance, instead of forcing *it* to surrender (as Aristotle here does by taking a characteristically tough and ingenious stand). In that case, one would, I think, be offering a golden opportunity to those who defend the doctrine of Platonic Ideas. Such philosophers might take transitive activity to be paradigmatic of activity as such and claim in consequence that activity is essentially incomplete. It would then not be difficult for these philosophers to declare every activity a reaching out for something beyond (*para*) itself, namely for a Platonic Idea. Thus even activities such as perceiving or thinking or basic biological functioning might come to seem aspects of a creature's striving towards, or imitating, the eternal, changeless, Platonic Idea of its kind. But this Idea, according to the Platonists, is in fact the substance, the substantializing principle, of the physical individuals which exist by trying to live up to it. These individuals are thus made substantial not from within by their own vital activities, but from something beyond themselves. It is this sort of view, I suggest, that Aristotle is trying to combat when he insists on the telic status of even transitive activity.

APPENDIX: ANALYSIS OF THE ARGUMENT OF *METAPHYSICS* 1050A4–B4

I. 1050a4–14 (*Alla mēn . . . theōrein*)

Statement of what is to be shown – that *energeia* is prior to *dunamis* in *ousia*, 4 – with two supporting reasons: (a) *prōton men hoti* . . . , 4–7, and (b) *kai hoti* . . . , 7–10.

[16] Aristotle emphasizes the incompleteness at *Ph.* 3.2.201b31–2; 8.5.257b8; *Metaph.* Θ.6.1048b18–22; *EN* 10.4.1174a19ff.

Reason (a) equates *energeia* with *eidos*; reason (b) equates it with the *telos* for the sake of which a *dunamis* is acquired or possessed. Examples of (a) at 5–7 (child is potentially male adult, etc.), of (b) at 10–14. The examples of (b) divide in two ways: (i) innate *dunameis* versus acquired ones; (ii) *dunameis* for intransitive activities (e.g. seeing, theorizing) versus *dunameis* for transitive activities (e.g. house building).

II. 1050a15–16 (*eti hē hulē . . . en tōi eidei estin*)

Generalized illustration of reason (a) above: matter is actually an *F* just when it is in the form of an *F*. (The formulation makes us focus on cases where the end brought about by the agent is a static hylomorphic compound such as a house.)

III. 1050a16–21 (*homoiōs de . . . ka'keinos*)

Two points: (i) something similar (sc. to the point made in II) holds where the end (*telos*, 17) brought about by the agent is an activity (sc. in something or someone else): for example teachers show their students *performing*; and (ii) something similar holds in the case of nature (*phusis*, 19; here the comparison is with technical cases).

[By now, it has been established that *energeia* is the *telos*, and the corresponding *dunamis* is for the sake of it.]

IV. 1050a21–3 (*to gar ergon telos, hē de energeia to ergon – diho kai t'ounoma legetai kata to ergon – kai sunteinei pros tēn entelecheian*)

Here is further support for the thesis that *energeia* is the *telos*: premise (1): the *ergon* (i.e. the typifying result) is *telos*; premise (2): the *energeia* is the *ergon*; ergo, the *energeia* is *telos*. Premise (2) is supported by the etymological connection between *energeia* and *ergon*. The impression that there is some kind of intimate connection between *energeia* and *telos* is further supported by the observation that the *energeia* strains towards fulfilment (*entelecheia*) – an observation which must be understood in the light of the etymological connection between *entelecheia* and *telos*.

V. 1050a23–b2 (*epei d'esti . . . zōē gar poia tis estin*)

(1) Activities are divided into intransitive (e.g. seeing) and transitive (e.g. house building), where the *ergon/telos* is something beyond the activity (23–7).

(2) It is asserted (2 i) that in the intransitive cases the activity *is* the *telos*; and (2 ii) that in the transitive cases the activity is the *telos* more than the *dunamis* is (27–8). (The thought motivating Aristotle's strong assertion of (2 i) is that absence of an *ergon/telos* beyond an activity might appear to indicate that this activity is not a *telos*. Aristotle is rejecting this appearance as false.)

(3) Then two grounds are given for (2 ii). They are (3 i) that the activity of house building is (not in *X* who builds houses, but) in the material being built up into a house, and (3 ii) that the activity of house building both comes-to-be and is simultaneously with the house (28–9).

(4) Now comes a comprehensive summary (with examples) of doctrine on the location of an activity: (4 i) where *X* is transitively active, the activity is (not in *X*, but) in the patient; and (4 ii) where *X* is intransitively active, the activity is in *X* (30–1050b2).

VI. 1050b2–3 (*hōste phaneron . . . estin*)

Intermediate conclusion: *ousia* and *eidos* are *energeia* (and not *dunamis*).

VII. 1050b3–4 (*kata te dē . . . dunameōs*)

Ultimate conclusion (from VI): *energeia* is prior in *ousia* to *dunamis*.

CHAPTER 10

Political community and the highest good

John M. Cooper

INTRODUCTION

The *Nicomachean Ethics* announces itself as a treatise on the highest human good, the "end" (*telos*) of human life – namely, *eudaimonia* or happiness. In the last chapter of the work (10.9) Aristotle makes it clear that the studies that he has by then carried out in investigating the human good leave his objectives in the treatise not yet completely achieved. The leading themes so far, he says in the opening lines of this chapter, have been the happy lives of philosophical contemplation and political leadership, the virtues (of character and thought), friendship, and pleasure. But in fact, he began the work by saying (1.1–2) that the study it contains is intended as a contribution to "political knowledge" (*politikē epistēmē*) or the political capacity or power (*dunamis*).[1] Its work will not be complete, he now says, in book 10, until a successful reader (or hearer) has been brought actually to possess that knowledge or power – political knowledge. By that he means the fully accomplished capacity for expert political engagement in affairs of state. Before the aim announced at the beginning of the *Ethics* can be achieved, then – that is, before we can fully define and explain in the right

I thank Danielle Macbeth for inviting me to speak at the November 5, 2006, conference at Haverford in honor of Aryeh Kosman, which was the occasion for the initial version of this chapter. I thank her and the audience at Haverford for a lively and helpful discussion, and particularly Aryeh Kosman and Amélie Rorty for questions and comments. I also thank Gabriel Richardson Lear and the other members of the University of Chicago workshop on ancient philosophy for their spirited and instructive discussion when I presented it there in February 2007. Finally, I thank Jonathan Beere for his written comments when I presented it at the annual Princeton Classical Philosophy Conference, December 1, 2007, and for his part in the lively discussion that followed. In preparing this final version I have tried to respond to the good suggestions and comments made on all these occasions, as well as to incorporate several improvements deriving from helpful discussions with Hendrik Lorenz. Knowing Allan Gotthelf's long-standing interest in Aristotle's, and Aristotelian, ethical theory, I am pleased to be able to offer this chapter in his honor.

[1] See 1094b10–11, where Aristotle says that his course of study (*methodos*) in the *Ethics* is in a certain way a political one; just before, at 1094a26, he says that this study is aimed at acquiring or conveying the knowledge (in outline) of the highest good and that the "knowledge or capacity" in question belongs to the "most controlling or most architectonic" science. He identifies that as "political" science.

sort of way the highest human good, or *eudaimonia* (I shall say more in just a moment about what this right sort of way is) – we need, as he puts it in *EN* 10.9 (1180a32ff.), to become expert in the establishment of good laws (*nomothetikē*) and good constitutions (*politeiai*, cf. 1181b14, 19, 21).

Now, one might certainly find this a surprising claim. As Aristotle himself is in no doubt, *eudaimonia* is a feature of the lives of individual persons. On his account, it is an activity, or a unified set of activities of individual persons. It is the active and devoted employment of the human virtues, organizing all one's other attachments and pursuits, in all the actions and activities of a person's mature lifetime. This commitment to the constant, active use of the human virtues determines one's overall preferences, one's choices and actions, thoughts and feelings, and all the pleasures taken in all the activities that make up one's life. But what does political knowledge – defined as expertise in the establishment of good laws and constitutions – have to do with defining and explaining this highest human good, and showing that it really *is* our highest good – this active employment by single individuals of their human virtues? One might easily enough see that these virtues might depend, in any number of ways, upon favorable political circumstances, including good laws and a good constitution, if any person is to employ the virtues in the way required in order to be fully happy in their life. Perhaps, too, if one wanted to bring happiness about for some group of other persons (not, or not just, oneself) one would need to know about good laws and constitutions, as necessary background conditions. So knowledge of politics certainly might reasonably be thought a valuable ancillary to knowing about the virtues and virtuous living.

But Aristotle goes further than that in book 10. He says that "we" (meaning his readers or hearers, including himself as studying alongside them as he lectures) need now to study laws and constitutions in order to be able, ourselves, to *act* in accordance with the virtues (1179b1–2) – that is, to attain happiness and live happy lives. We will not be *able* to attain the virtues ourselves and live happily through making them the organizing principles of one's life, until we have acquired political science – the expert knowledge of laws and constitutions. So Aristotle seems to hold that in order to fully grasp, or define and explain – in the right way, at any rate – what the highest good for a human being is – this is the whole treatise's subject of study – we must conceive the happy activity itself somehow in political terms, terms investigated and explained in the *Politics*, and not simply in individual ones at all. It is only by understanding it in that political way, he thinks, that one will be able to act fully and properly according to the virtues. In this paper I attempt to work out and explain what Aristotle intends by this essentially "political" orientation of the activity of virtue, that is, of the activity constituting human happiness.

PRACTICAL KNOWLEDGE/UNDERSTANDING

Let us begin by considering the reasons Aristotle gives at the end of the *Ethics* for holding, as we have seen he does, that the controlling intention (*proairesis*) in the treatise has not yet been completely fulfilled – even though as we will see they unfortunately raise more questions than they settle. The first thing Aristotle does is to remind us that where it is a question of things to be done (*ta prakta*) – actions and activities – our goal in our studies is not to develop theories (*theōrēsai*) and to know (in that "contemplative" or theoretical way) about the various points concerned. Rather, our purpose is to come to act, ourselves, in all the relevant ways, by possessing and exercising the virtues (1179a35–b4). This echoes what he said back in book 2, as he began his discussion of the ethical virtues: "The present undertaking (*pragmateia*) is not for the sake of developing theories as our other ones are (for we are not inquiring so as to know what virtue is, but so as to become good people). So we must inquire into the subjects relating to action, that is, to how one should act" (2.2.1103b26–8). This itself echoes and expands what Aristotle said already in 1.3.1095a5–6: "the goal" in political studies "is not knowing but acting." As these earlier passages from books 1 and 2 imply, Aristotle thinks that it is a necessary, or anyhow an especially effective, means to becoming a good person (that is, a *fully* good one, one who actually does fully possess the virtues and lives constantly on their basis) to engage for oneself in the philosophical thinking and argument involved in studying ethics with him, in the progression of discussions that make up the *Ethics*.[2] Following his lead in these three passages of books 10, 2, and 1, we can put the difference he notes between moral philosophy and studies in metaphysics or philosophy of nature by saying that the understanding sought in moral philosophy is not a theoretical (or "contemplative"), but a practical, understanding – one that immediately or directly leads to one's living virtuously. It is an understanding, which he thinks qualifies as a special *sort* of understanding, because of this immediate or direct connection to action. The purely *theoretical* studies in physics and

[2] Being fully good means possessing all the human virtues for action, including practical wisdom. Aristotle implies in the 2.2 passage that, even if it might be possible (for all he says there) to become practically wise (and therefore fully good) without philosophical study, the study of ethics that he is engaging his readers in aims at that goal. See *EN* 6.13.1145a1, referred to below.

metaphysics do not have such a connection. Nor indeed would a similar, purely contemplative study of human action itself, or of the virtues and human happiness.

We can begin to understand why Aristotle insists that the understanding sought in moral philosophy must be sharply distinguished from the understanding of other matters if we bear in mind that, as he frequently says, actions are always done for the sake of some good – in fact, some *human* good (as he explains in I.6). The understanding being pursued in this case, but not the others, is of what is good for a human being, where what is good is *understood* as being good for them. One cannot attain, Aristotle thinks, an understanding of such good things as *being* of value for us just by knowing in a non-committal, "theoretical," way what they are, or even what is good about them. Someone might perhaps be able to go through all the sound arguments and understand and explain, and defend with full articulateness, on their own behalf (and not just by repeating what some teacher has said), and thus actually grasp, all the reasons why, in fact, virtue and virtuous action are supremely good for a human being (i.e. for themselves, among others). But if they were left indifferent by knowing these reasons, they could not properly be said to *understand* the goodness of virtue. Anything entitled to be called an *understanding* of goods, *as* goods, including virtue itself, must include a motivation for becoming and being virtuous. Indeed, as we can see from what Aristotle says in the passages cited above, it must lead to constant and reliable action in accordance with virtue, on the basis of that motivation. For our immediate purposes, however, we need to focus first upon his claim that this understanding does, necessarily, involve *some* motivation toward achieving or realizing any good so understood, that is, understood as *being* good.

It is important to see that for Aristotle this practical understanding of virtue really is *understanding* of it as good, and nothing more. It is a full, explicit, articulable grasp of what is good about virtue, about how that good relates as a value to other things also similarly grasped as good. This grasp, all by itself, he believes, moves us to embrace virtuous activity as our highest good, and to make it the organizing and controlling center of our whole lives. For Aristotle, there is no feeling of attraction to virtuous activity, as something *separate* from the understanding, that needs to be added, in order for the motivation provided directly in this understanding to be present and at work in one's psyche.[3] This is a point that is often missed

[3] This is not to deny that in virtuous people there will also be some additional feeling of attraction for virtuous activity, something emotional in character, as we could say; my point is only that for

or misunderstood, since it flies in the face of modern philosophical and psychological assumptions about the separation between reason or rational understanding and desire or motivation. So it is worth both emphasizing and dwelling on briefly.

Aristotle holds that to think assertively of something as being good (for oneself) *is* to be moved thereby toward it: this being moved is part of or an immediate effect of that thought itself (it is what he calls a "rational desire," or "wish," a *boulēsis*).[4] Because of the essential connection to motivation implied in the very act of understanding (and so, assertively thinking of) something good *as* good for oneself as a human being, this is a special *sort* of understanding. This understanding (when fully accomplished) is what Aristotle identifies as one of the "intellectual" virtues, or virtues of thought (*dianoia*, cf. 1103a5), as opposed to the virtues of habituated states of feeling (*ēthos*) – the ethical or moral virtues. When this understanding is fully accomplished, its motivations are so deep and strong that Aristotle holds that it not only motivates one who has it to act virtuously (as I mentioned above), but is *sufficient* to bring it about that they do act that way, constantly. This understanding comes in degrees, of course. Still, any one who has it even less than completely is motivated by it toward virtuous action. Having the motivation is compatible with not acting on it, because, of course, for Aristotle there are other motivations as well, which might lead the agent to act on them instead. But when one understands fully that and why virtuous action is the *highest* human good, this knowledge is in fact the virtue of *phronēsis* or practical wisdom, and it guarantees right action. By having acquired that complete understanding we become fully or simply or without qualification good and virtuous, on Aristotle's account – *haplōs agathos* (*EN* 6.13.1145a1). So, when Aristotle says in *EN* 10.9

Aristotle practical understanding of values, as values, in and of itself provides its own, separate, sort of motivating push or pull toward virtuous activity. On the additional feeling, see below, on the "love of the fine" about which Aristotle speaks in this connection. This love is just such an emotional attraction, and all virtuous people experience it, in addition to the purely rational motivation provided by the understanding itself in grasping the value of virtuous activity.

4 See "Some Remarks on Aristotle's Moral Psychology," in Cooper 1999. As for texts of Aristotle, see especially *MA* 7.701a7–14, where one must note carefully that Aristotle is saying that *whenever*, first, one thinks as assertion one is making in thinking them, and thinks them together as one assertive thought, the two premises of a syllogism, whether in theoretical or in practical thinking, then one psychologically *must* (in the one case) go on to think (assert) the conclusion, and (in the other case) to *act*. (One must add, in the latter case, that action might be prevented by a countervailing motivation, such as excessive appetite or anger.) His claim is that rational beings are subject to this pair of psychological necessities. In the practical case the thought of the two premises, together as one, constitutes what Aristotle only later, 701b1, calls a *boulēsis*. This special motivation belongs to reason itself: simply holding, on some basis of thought, that something is good, and thinking of it *as* being good, includes a motivation toward having or achieving that good thing.

that our purpose in going through the investigations of the *Nicomachean Ethics* is to come to act in certain ways, that is virtuous ones, he is reminding us that the study just being completed is aimed at making us fully good people by giving us this sought-for practical understanding of virtue as our highest good.

Now, Aristotle notoriously insists that no one is to take part in the philosophical study of ethics and politics without having first acquired good habits of feeling through their earlier upbringing and education (1.3.1094b28–1095a11 and 4.1095a30–b13). Having achieved such habits of feeling enables people to go forward, if they are sufficiently gifted intellectually, so as to grasp the philosophical principles that ground the further virtue of practical wisdom. In fact, grasping those principles turns those early habits into fully virtuous states of character. These initial habits are sufficient, but sufficient *only*, to give people an intuitive attraction to proper behavior and to the values it serves, and an intuitive dislike of the opposites. This intuitive attraction is the love of *to kalon* ("the fine," as recent translators have inadequately begun to translate it),[5] that Aristotle makes a predisposition necessary for ever becoming virtuous. It is also the characteristic motivation of morally virtuous people insofar as they are, precisely, possessed of virtue of *character*, that is, possessed of good states of habituated feeling once those have been fully grounded in practical wisdom. Without such preliminary habits and intuitive feelings, one is not open to grasping the reasons why the one sort of behavior (virtuous) is such a good thing for oneself, and the other (vicious) so bad: one will not listen if someone tries to explain these reasons, or will not understand if one does (10.9.1179b23–8). Only by having these habituated feelings is one now ready to pay attention to what reason and philosophy have to say. Thereby one can acquire the new, purely reason-based motivation for acting virtuously that I have described. In feeling the attraction to the fine, a young person is not led, as people not brought up to have good habits of feeling are, simply by the passions (especially those related to untutored immediate pleasure and dislike) that proper behavior places restrictions on. Their love of the fine provides a counterweight to the passions. Hence, as Aristotle says, acquiring the knowledge of philosophical ethics and politics,

[5] Old translations tend to use "the noble" or "honorable" or (morally) "beautiful." "Fine" probably strikes a better note in British ears (the recent translators have mostly been British) than in American. But there seems to be no translation that captures well the force and nuances of the Greek term. It connotes balance, harmony, order, and the attractiveness inherent in anything with those qualities for any rational being. If our modern understanding of beauty better highlighted that basis of attractiveness one might opt for the translation "beautiful." So, though I greatly dislike its archaic overtone, I will continue to use "the fine."

through attending lectures on ethics, would be of use to such a person. People already equipped with the love of the *kalon* are in a position to be improved by this knowledge. These studies would advance them toward a full understanding – a *practical* understanding as I have just explained it, involving reasoned motivations – of what is good and what is bad for human beings, quite generally, and so of the consummate value of the virtues in giving shape to a human life, including, of course, one's own. In doing so, it would sharpen and deepen, and expand the scope of, those preliminary habits of feeling, by bringing to light new values, or new aspects of old ones, to be brought within the purview of one's emotional attachments and feelings of attraction or aversion.

DECENCY VERSUS FULL VIRTUE

It is very important to realize, as commentators often do not, or tend to forget, that neophytes who first come to the study of ethics equipped with the love of the fine are, however, not good and virtuous people already – despite their habitual practice of (more or less) virtuous behavior, and their intuitive love of the values that such behavior constitutes and promotes. For Aristotle, full virtue requires much more than simply firmly established habits of non-rational feelings and desires falling within a correct, intermediate range: it requires rationally developed understanding about the whole realm of human values. This understanding, then, both confirms as basically correct and, with reason's own innate desire for the good, directs the habituated non-rational feelings and desires so that they do conform fully to the rationally determined standards for their correctness. These standards, in turn, become the basis for corrections to, and extensions of, the habits of feeling one has learned previously; the philosophical understanding of human values identifies subtle distinctions and somewhat new perspectives from which to enrich and reshape the already quite good results of childhood training toward the virtues. Thus, those who come to the study of ethics/politics are only basically decent, young adult, but still somewhat unformed people.[6] Certainly, they are not, strictly speaking, morally virtuous yet. Indeed, they do not yet have well-settled characters at all. However, having made a good start through their upbringing and their

[6] At I.3.1095a2–4 Aristotle emphasizes that adolescent boys (who might be ready for other philosophical studies) are not suitable students of ethics and politics, in part because this study presupposes considerable experience of life. So those who come to study ethics with him are past the stage of adolescence and have begun to have, as young adults, the additional qualifications in experience of life that Aristotle requires for the effective study of philosophical ethics. Only adult people have all the qualifications he requires.

experience of life so far, they may come to possess well-settled respectable characters, and so to be decent fully mature people, even *without* engaging in philosophical study of ethics and politics. A basically decent twenty-year-old with no philosophy could advance beyond this initial, still somewhat unformed moral state, so as both to refine the rather crude habits acquired from their upbringing and, through experience and reflection, to come to grasp, to some extent, the true system of values that philosophy establishes as true.

In that way their reason, even though untutored by the discipline of philosophy, would add its correctly, though incompletely, informed support for and direction of their habituated feelings and desires, by way of its "wishes." They would thus finally acquire, as mature adults, the fully settled character of decent, respectable people, who lead good but wholly ordinary lives, not enlightened by philosophy.[7] But Aristotle's hearers in his lectures are presumed to have incipient characters so disposed, and their intelligence is such, that they *can* acquire the understanding that philosophy provides. They can thereby become *more* than mature decent people. As the copestone to the development of their merely instinctual feelings into that condition of settled, fully adult decency, they can add, through philosophy, a cultivated and informed, argued and articulate, grasp of the whole realm of human values. Their practical reason, and its special motivations, can be brought in to clarify, adjust, and support their mature, merely intuitive feelings. Hence, when Aristotle says in the passage quoted above from book 2 that our undertaking in the philosophical study of the virtues aims at our becoming good people, he is taking for granted that anyone engaging in these studies is already a basically decent young adult, destined to become a decent fully mature person in the normal course of events. He is saying that by learning what philosophy has to teach us about ethics (and politics) we acquire the virtue of practical wisdom and become more than totally decent people. That is, we become *fully* good. Thereby, we come to live the fully and perfectly happy life.

Accordingly, when Aristotle says at the end of the *Ethics* that our goal is not yet completely achieved, he means that, because from the outset

7 In this sketch of "decent" people I am offering an expanded interpretation of what Aristotle understands the people to be like to whom he refers at *EN* 10.9.1179b5 and not infrequently elsewhere as *epieikeis* (see below, pp. 220ff.). Sometimes he calls them *metrioi* (4.1.1121b6; see also 1125b13, 5.12.1136b20, and *EE* 2.5.1222a34); and often the context suggests that even when speaking of people as *spoudaioi* he means not (as he sometimes does) the fully virtuous – possessed of virtues both of character and practical intellect – but these lesser, more ordinary, good people possessed of that basic decency that approximates to the virtues of character alone. As my discussion below, I hope, shows, this category of good moral agents plays a much larger role in Aristotle's ethical and political theory than commentators have generally recognized.

of the work (being already basically decent persons, with good habitual ways of feeling about proper behavior and the values it serves) our aim has been to become fully good ourselves, we need something beyond the studies already concluded (as he says) in outline form (*tois tupois*). At first Aristotle says, without mentioning specifically any further *studies*, simply (as I reported above) that "we need to try to possess and use the virtues – or however else we may become good people" (1179b3–4). One can indeed readily understand that, even if our philosophical studies have been carried out as efforts in acquiring practical knowledge, we might need time and effort – further practice – to entrench our newly established philosophical understanding in our minds and to bring it to bear in further habituating our feelings and in bringing them fully in line with our reasoned convictions as we act. Moreover, as I have mentioned, we can suppose that the philosophical understanding achieved through the practical knowledge that our studies so far have given us could very well lead to revisions in our practical attitudes of emotion and appetitive desire, as those existed in us when we began our studies. Our philosophical understanding of the various sorts of goods there are, and of the role of emotions and appetitive desires in our pursuit and use of them, as well as our grasp of just which states of these non-rational feelings are "fine" (*kalon*) – and just what it *means* for them to have that character – will be clarified and deepened. This presumably would involve some revisions in our views of precisely or approximately just *which* feelings and actions really are the correct, and the "fine," ones to feel and do habitually on various occasions and in various recurrent (or not recurrent) situations. So, to be sure, there might be further effort, of these two closely related sorts, to be made in "trying to possess and use the virtues," even after we have thoroughly learned and become persuaded, as primarily an intellectual matter, of the correctness of Aristotle's analyses and arguments in the *Nicomachean Ethics* about virtue and its place in a well-lived life. We need still to work at making what we have learned fully effective in our lives, before we can become fully good people ourselves.

NICOMACHEAN ETHICS–POLITICS AS HANDBOOK FOR POLITICAL LEADERS

But that is not at all how Aristotle continues. He launches immediately into a protracted discussion of how to make people decent (*epieikeis*), with his focus, apparently (but, I will argue, misleadingly), not on what is needed now, as a further something, to make *us*, who have been studying moral philosophy with him, fully good, but on what is needed to make *most*, that is ordinary people, at least decent in their behavior and in their lives. And that then leads him to specify further subjects of study that we need to undertake. He begins like this:

> Now, if arguments (*hoi logoi* – discourses of one sort or another) were sufficient by themselves to make people decent they'd win many fat fees, and rightly so, as Theognis says; arguments would be what had to be provided.[8] But in fact, though they do appear strong enough to urge on and motivate young people who value being in charge of their own actions,[9] and to make a character that's well born and truly loves what is fine be possessed by virtue, they appear unable to urge most ordinary people on toward being refined and good (*pros kalokagathian*). (1179b4–10)

Now, in fact, we have heard all this before, in the passages of book I that I summarized above, concerning the need for a good upbringing before beginning philosophical studies of ethics. Hearing this again now is disconcerting, in two ways. For one thing, Aristotle grants here that arguments (such as his own treatise is full of) do have the strength to urge on and motivate people like us, his authorized hearers, and (as he says explicitly here) to make our characters be possessed by virtue. So if we have been attending properly to the arguments of the previous ten books, would we not now already *be* possessed of virtue – or, at least, not in need of further *study*? We might still need, if anything, to get practice in (fully) virtuous thinking and acting, in a fairly wide variety of circumstances, along the lines I just indicated, with adjustments in our habituated states of moral virtue, so as to feel emotionally the clarification and deepened insight that our increased practical understanding has effected for us. But why would we need any further *study*, any study of additional topics? Yet, as I indicated at the outset, in fact Aristotle is building up in this chapter to telling us we need to study laws and constitutions.

On the other hand, Aristotle's proposal of such further studies certainly is relevant to the task of dealing with a whole population and doing what one can to improve them. It may well be that, as he says here, most people are not going to be improved at all in their behavior by arguments and discourses (that is, by the only sort of thing that philosophy, even practical philosophy, provides). But perhaps they could be improved by having good coercive laws to guide them: this is what Aristotle's discussion of laws a bit

[8] I follow Rowe here (Broadie and Rowe 2002) in taking *toutous* to refer not to *misthous* (fees), as e.g. Ross and Irwin do, but to *logoi* at the beginning of the sentence. Taken the other way, the clause does little more than just repeat what was already said with *dikaiōs* (justly). Taking it Rowe's way makes better sense, in context, too.

[9] I.e. the *eleutherioi*, lit. the free: I cannot think of a single English adjective that could capture what I take the meaning here to be. Ross's "generous-minded" and Irwin's and Rowe's "civilized" are no good.

further on makes clear. So that could be one reason why we readers and hearers of his *Ethics* might need to engage in further study – *if* we assume (I shall say more about this in a moment) that *one* thing Aristotle has been aiming at in his treatise, and one thing we have been studying with him for, *is* to learn how to improve ordinary people's lives. Still, how is that further study relevant, as the preceding context clearly implies Aristotle thinks it is, to the needs for *self*-improvement of the persons Aristotle is speaking to in writing the treatise?

These are puzzling questions. Alas, I do not think Aristotle's subsequent discussion in this final chapter of the *Ethics*, either implicitly or explicitly, does clearly give us the materials we need to resolve them. I do think, however, that we can work out a resolution if, as I shall do below (pp. 228ff.), we bring into our discussion fundamental aspects of Aristotle's theory of the political community in his *Politics*. In any event, what happens in this last chapter of the *Ethics* is this. Aristotle first raises the question of what he and *we*, his authorized, successful students, need, so as to complete our project of becoming fully good people through our philosophical studies. He then introduces, without explication, a second purpose that he now presupposes we and he have had in mind all along. This is the aim of using our practical knowledge and good characters in offering leadership in helping other, ordinary people to become at least decent and live good lives at the level of decency, a lower level than that of full virtue. This aim has not in fact been clearly announced heretofore, as we have seen the first purpose, of self-improvement, has been, in book 2. But it might be thought to be implied already in the opening discussion in book 1 which I began by referring to. 1.1–2, arguing for the conclusion that our enterprise in the treatise is to be classified as belonging to political knowledge, conclude as follows:

> For even if the good is the same for a single person and a city, the good of a city is evidently a greater and more final or end-like thing both to achieve and to preserve. While it is gratifying to do this even for a single person alone, to do it for a people and for cities is a finer and more godlike thing. So, then, our course of study seeks these things, belonging as it does in a certain way to political knowledge. (1094b7–11)

Here Aristotle clearly suggests[10] that the *Ethics* has as at least one of its ultimate goals to help its readers come to know (as an element of practical,

[10] He would say this explicitly if, with Irwin in a bracketed explanation in his translation of "these things" in the last sentence, we glossed them as "these goods, for an individual and for a community." But Burnet (1900) seems more right to take "these things" to mean "to secure and preserve the good for man." It only results from special facts about what the good for a human being – as an

not theoretical, knowledge) not only the highest good for a single individual (oneself, he obviously means) but (even more) the much larger and more complex highest good of a whole people or city.[11]

Accordingly, in the last chapter of the *Ethics*, having begun his response to the question whether our work in studying the topics of the treatise is now complete by referring to the first of his and his readers' purposes (self-improvement), Aristotle shifts abruptly and confusingly to the second (achieving the good of whole cities). From that point onwards, he diverges into a discussion of the need for an appropriate civic constitution and system of laws if most people are to become at least decent and live decent lives.[12] Such a constitution and laws are needed, in the first instance, he goes

individual – is, and how it is to be achieved and preserved, that (as I argue below) in knowing it one also has to know the highest good for a people and for cities. Hence the most Aristotle says explicitly here is that we need to learn what the human good is, and how to secure and preserve it, through our studies in ethics to follow; he leaves it unspecified whether we will do this for the sake of securing our own good individually or (also) that of cities and peoples. It is true, of course, that by describing the latter goal as finer and more godlike he suggests that we should be pursuing the latter goal as well as the former one in our study of ethics. But he does not say that outright.

11 We are indebted to R. Bodéus (see *Le philosophe et la cité*, 1982; English translation, 1993) for forcefully drawing attention to this political dimension of Aristotle's *Nicomachean Ethics*. According to Bodéus, Aristotle gave these lectures exclusively to persons who came to hear them as ambitious aspirants to political leadership in their cities, not as aspiring philosophers, or laymen wishing to get what benefit they could from philosophical understanding for leading their private lives; Aristotle's lectures on *Ethics* teach them what they need to know about human happiness (and related topics) as a first step toward learning, further, through Aristotelian lectures on political systems and systems of laws, how to design and direct political institutions with a view to bringing about good order in their cities. This benefits the ordinary citizens by making their lives as happy as is possible for them, but equally or more important, it prevents them from interfering inappropriately and damagingly in the philosophers' freedom to do their esoteric philosophical work and undermining the civic support they need for living their even happier lives. Bodéus's unfortunately one-sided account of *EN* 10.9 (see his chapter 2, pp. 47–68 of the Eng. trans.), on which much of his overall view rests, overlooks altogether the implications of the opening of the chapter, where Aristotle speaks of what "we" need to do to become good people (he nowhere in his book even alludes to the opening lines of *EN* 2.2). He also naively misunderstands Aristotle's theory as assigning the ethical virtues to people solely and sufficiently on the basis of habituation in decent ways of feeling and acting (see pp. 2, 51, among other places). In fact, as I have emphasized already and will emphasize still more below, for Aristotle one cannot become a fully good person without a developed understanding of basic truths about human nature and communal human life. Bodéus misses altogether the intended role of Aristotle's lectures in both *Ethics* and *Politics* in educating their hearers (and readers) for that most fully realized possession of the ethical virtues. It is a serious error to suppose, as he does, that the education of students who intend to function as (semi-)professional political leaders is Aristotle's sole or even his leading purpose in his lectures. It is one, but only one, of two principal ones. (See further in my text below.)

12 One of the provisions of a good constitution will be systems of laws specifying and encouraging decent, and discouraging, even punishing, indecent behavior. However other provisions will establish philosophy and philosophical instruction in the city – at any rate, I take this to be a somewhat under-explicit provision of the city that "satisfies our dreams" sketched in books 7 and 8 of the *Politics*. Hence, Aristotle may be suggesting in *EN* 10.9 a further purpose to be achieved by having his students learn about constitutions. At least some of them, by going on after completing

on to argue, because the use or threat of painful punishment is necessary to bring most young people to accept and perform the actions they must perform regularly if they are to be habituated to decent practices, and thereby gradually to establish in themselves a decent outlook on life as adults. Most people when young (unlike the few whose natural characters, he says, make them love what is fine)[13] cannot be moved by shame, but only by fear. They pursue pretty much only the pleasures of gratifying their passions, and only the threat of countervailing immediate pains can deter them from pursuing them when it is not decent to do so (1179b11–16).

It is true, of course, as Aristotle says, that at this early stage of life a person's upbringing is largely in the hands of their parents and others in the household. But a parent's directives, unless backed up by fitting into legally established and required practices (and seen by the child to be so backed up), can lack the necessitating force required to bring the child into line. Children, as Aristotle says, tend to hate people who oppose their impulses, and they resent and rebel against what could reasonably seem to them to be the merely arbitrary orders of some single individual, even a parent (whose manifest goodwill and natural connection from birth might have a countervailing effect; see 1180b4–7). When backed up by the law, however, the parent's directives obtain a different aura. The universality and wide acceptance of laws suggest, even to children, that they are based on good reasons. Laws present themselves as imposing a correct orderly regimen, one that there are good reasons to accept, even if those reasons may sometimes be less than fully evident to those subject to them. So children will not feel a parent's directives when backed by laws as arbitrary and burdensome but will accept them as resting on good reasons and imposing proper standards of good order. They will act decently and even forego the immediate gratification of their passions *willingly*, as is necessary if they are to develop good habits of feeling and action and acquire some sort of reasoned intelligence and sense of good order of their own, so as to become decent or (at a popular and non-philosophical level) what we call good people. (For all this, see 1180a14–24.)

their studies to function as active political leaders, will use their political knowledge in instituting philosophical study in their cities. They will thus provide the conditions needed if people like some others, and the most intellectually gifted, of their fellow students can be raised and educated so as to live not only fully virtuous lives, but completely happy ones – not just happiest in the second rank, but happiest, period. On the distinction between second happiest and absolutely happiest lives, see further below, pp. 225–7.

[13] Among these few, of course, are to be found all Aristotle's candidates for the effective study of ethics – once they have acquired sufficient experience of life.

Moreover, Aristotle argues (1180a1–5), it is not enough for most young people to receive a caring and correct upbringing. Having reached adulthood they also need to maintain the same habits of feeling and action as they learned as children. That, too, Aristotle says, requires law and the threat of punishment, now without the intervention of parental directives as their intermediary. Most people, even when well brought up, are always more moved by fear and the threat of punishment than by the power of good reasons or by their sense of what is fine, when those oppose their immediate gratification. Hence anyone out to improve most people's lives needs to know about the correct system of laws for people as adults to be living under, and in general about the principles of politics. They need, in fact, to equip themselves so that they would at least be *qualified* to be political leaders – active politicians, concerned, through their possession of political knowledge, with the establishment, administration, and preservation of good laws in their own or others' cities.

If, then, one of Aristotle's purposes in the *Ethics* is to prepare intellectually well-endowed and well-brought up young adults so that they will be able to provide the highest human good (so far as possible) for a whole city, it is clear that the course of study completed in the *Ethics* itself has not accomplished all its goals. From the beginning we were told that the *Nicomachean Ethics* is intended to convey political knowledge to its readers, as an item not of "contemplation" but practice. Part of this it has conveyed, namely the basic account of the human good, conceived so far, basically, in terms of single persons' lives led as separate individuals. But readers need also to study and learn about constitutions and systems of laws. They need to know which ones are required for providing the human good for a given whole city (so far as possible – that is, by making all the citizens at least decent people). And they need to know how one brings those laws into existence or preserves them, through political engagement and activities. Accordingly, Aristotle concludes 10.9, and the whole of the *Ethics*, by arguing (1180b28–1181b23), in effect, that only a fully systematic account, based on general first principles of politics and ethics, of just the sort we find in his own *Politics*, can suffice for completing the course of study initiated at the beginning of the *Ethics*.

NICOMACHEAN ETHICS–POLITICS AS AID TO SELF-IMPROVEMENT

Now, as I have said, all this concerns the achievement only of the second of the two potential objectives for the study of ethics that Aristotle distinguishes in the passage I quoted from *EN* 1.2, namely the highest human good for a people and cities. Might we, however, also find a connection between knowledge of politics and the first objective, the highest good of a single person (oneself)? One possibility immediately suggests itself. In the chapters of book 10 immediately preceding 10.9 Aristotle has distinguished two lives as being happiest ones (the second being happiest "in the second rank"). These are the life devoted to contemplative knowledge of the best kind as its highest goal, and the life devoted instead to the exercise of the virtues of character, which involve practical knowledge and practical thought as well as habituated feelings but do not in themselves involve purely theoretical, namely contemplative, thinking (1178a6–22). Though Aristotle does not explicitly say so there, it seems probable that with this second happy life he is thinking of the life of the political leader, actively engaged in the political direction of his city.[14] This is what, near the beginning of the treatise, he calls "the political life" and contrasts with the theoretical one.[15] If this is right, then in speaking in 10.8 of a life of "the other virtues" as "happiest" (in the second rank), he is not referring to the life of a fully virtuous private citizen who does their political duty but keeps out of the political limelight. Surely, he does not think that the virtuous life of someone who kept out of active politics would *not* be a happy one; still, he does apparently think that the virtuous politician's life is a supremely happy one of the same sort, devoted to the exercise of the virtues of character and practical intellect as *its* actually achieved highest goal.[16] Only it should count as second happi*est*. And, as T. H. Irwin has argued at length, there are indications in the *Politics* itself that this is Aristotle's view.[17] Apparently he thinks that what is good about the exercise of the virtues of character and practical intellect is most fully realized only in the context of an active life of political leadership. In the political activities of this life these virtues receive their widest scope and are directed at the grandest of morally good goals – not just one's own, but at the same time and in the same activities, all one's fellow citizens' happiness as well, through their coming to live constantly, or nearly constantly, fully decent lives.

[14] See Lear 2004, 177ff.

[15] In 1.5 Aristotle distinguishes three conceptions of the happy life as particularly worthy of note (see 1095b17–19): the life of gross pleasure (which is worthy of note only because it is the way most people live), the contemplative life, and the life thought to be devoted to public honor and esteem, the "political life" led by political leaders (*ho politikos bios*). In that chapter Aristotle drops the life of pleasure from consideration. The two surviving ones resurface in book 10.7–8.

[16] Or, more precisely, as the highest activity actually achieved in it. Aristotle seems to think that even in this second happiest life excellent contemplation serves somehow as a goal of and for the practically virtuous activities it is centered round, though an unachieved one. See Lear 2004 and "Plato and Aristotle on 'Finality' and '(Self-)Sufficiency'," in Cooper 2004.

[17] See Irwin 1990.

If so, then we could, with a little work, find a connection between the first aspect of his project in the *Nicomachean Ethics* and Aristotle's insistence in *EN* 10.9 that the aims of that treatise will not be finally achieved until we have learned the principles of politics. If the life aimed at practical virtue as the highest good that one is capable of achieving is most fully and richly accomplished in the political life, then Aristotle's young adult students do need to have the knowledge of how to exercise their virtues as political leaders. And they will not know how to do that without learning the principles of politics. Thereby, or therein, they will acquire knowledge of laws and constitutions as the culmination and completion of the practical knowledge (practical wisdom) that they need in order to lead fully virtuous lives devoted to practically virtuous activity as their highest achieved good. It is only with that knowledge that a completely realized virtuous life of that sort, one with the widest and finest scope for the deployment of the practical virtues, can be shaped and led.[18] Thus, if we consider solely the aim in studying the *Ethics* of becoming good oneself and thereby living happily, Aristotle's hearers/readers need to continue their studies by listening to lectures about legislation and constitutions, because the further knowledge to be acquired therein is needed in order to be not only good and virtuous, but to exercise one's virtues with their widest scope. Thereby one will live, not just a happy life, but the (second) happiest one possible.

[18] See Aristotle's discussion of political knowledge in *EN* 6.8, starting with the assertion that "political science and practical wisdom are the same state of mind, but their being is not the same" (1141b23–4). Their being differs, Aristotle explains, because we use the name "political knowledge/science" for the same state of mind "as it relates to the *polis*," while as it "relates to oneself as an individual" it is given the name "practical wisdom." As it relates to the *polis*, he goes on to say, it is knowledge of legislation and political administration; as it relates to oneself as an individual it includes knowledge of household management. Clearly, on Aristotle's view, one cannot have either political knowledge or practical wisdom without having both: they are precisely the same knowledge, knowledge of (so to speak) the same body of knowledge. Strictly speaking, according to what Aristotle says in 6.8, one will call the knowledge being exercised "political knowledge" only when it is being employed in the way an expert active politician does, for achieving the good of cities through providing the means for as good and happy a life for its citizens as is possible; whereas, for a person who regularly applies that same knowledgeable state of mind only in living their individual life, the correct name to use is "practical wisdom." However, as I explain below, the identity of the underlying state of mind means that the knowledge being used to direct one's own life (whether as a fully virtuous political leader or a virtuous private citizen) makes essential use of the political orientation given in this passage for this body of knowledge. Aristotle here declares that body of knowledge to be the intellectual content of that practical understanding which, as we have seen, he aims to convey (but only in part) in his lectures on ethics. From what Aristotle says in this chapter it is manifest that if he intends the successful, approved hearers of his complete series of lectures on politics (what he gives us in the *Ethics* plus the *Politics*) to acquire political knowledge thereby (as he plainly and explicitly does), he equally intends that they acquire (completed) practical wisdom by the same means. This supports my claim above that the *Ethics* aims to help us become fully good by providing, in the content of its lectures, the essential basis of that practical understanding that constitutes the virtue of practical wisdom.

However, by itself this suggestion is not satisfactory. If this were all Aristotle had in mind in saying that we now need to learn about constitutions and laws, one would expect him not to have been so indirect about it. In *EN* 10.9 he could easily and naturally have built explicitly upon his argument in the immediately previous chapter. He could have said that since, as he has just argued, the happiest of lives devoted to the virtues of character and practical intellect is the virtuous politician's life, the project of helping his readers/hearers learn through their philosophical studies how to be *fully* good and happy is necessarily incomplete until he helps them acquire the knowledge needed for political leadership. Since he does not explicitly argue that way, I have had to fill in a considerable amount of background in order to construct this interpretation. Moreover, there is something unsatisfactory about this interpretation, taken on its own. It speaks as if for Aristotle one can distinguish some (non-political) knowledge of the virtues needed for living virtuously as an ordinary citizen, from the added political knowledge needed in order to live the more fully realized virtuous life of the virtuous political leader. It implies that the added knowledge is needed solely for enabling that fuller realization of virtue, and not at all for living the less happy life of virtue led by a virtuous private citizen. That sells Aristotle's claims for the political character of virtue very much too short (and see n. 18). So we should feel encouraged to look for further and different background for linking Aristotle's claims in *EN* 10.9 concerning the need for knowledge of politics (in a narrow sense) if his readers are to become fully good and live happy lives.

We can find what we need if, as I suggested above, we bring into our discussion fundamental aspects of Aristotle's account of the political community in the *Politics*.[19] When we do that, I will argue, we can see that on Aristotle's fully developed theory of the virtues of character and practical intellect, each and every exercise of them, if they are to be properly exercised at all, requires an orientation not just to one's own happiness (in exercising the virtues) but to the happiness (the virtuous living) of the others with whom one shares life in one's political community. On this view, virtuous activity when fully realized (even by someone leading a private, not

[19] Aristotle could hardly, of course, have done this explicitly himself in writing the last chapter of the *Nicomachean Ethics*. That would be to presuppose crucial results of the studies he is only suggesting the further need for. This fact must explain Aristotle's indirection in drawing on his views (*EN* 10.8) about the political life as second happiest one in justifying his claim that we need to study legislation and constitutions in order finally to become fully good people ourselves. It also allows us, his interpreters, quite reasonably to draw on the theories of the *Politics* as what must, in addition, have been in his mind in making that claim. He was thinking about both connections.

politically fully active life) has to be a communal undertaking, something engaged in by each virtuous (or even decent) person as his or her part of the single activity, engaged in in common with one's fellow citizens, of living according to the requirements of the virtues – that is, living that way as a shared, and mutually supportive, *common* way of living. This is something they all do together, in a sense that I will explain below. Whether in the context of active political leadership or not, therefore, each fully virtuous act, in being properly done, must be conceived as a contribution also to the virtuous life and happiness of the whole community, the community of all the virtuous or at least decent people in one's city. Each virtuous act, whether of a decent or a fully virtuous person, contributes, and is conceived by them as they do it as contributing, to the happiness of oneself together with all the others with whom one lives a good life in common. For this reason, even if we take up moral philosophy solely with a view to becoming good and virtuous individual people, and living happily as individuals or in our circle of family and personal friends, we nonetheless need to learn the principles of politics. Those are the principles for understanding, in a fully practical way, such a communal life of virtuous activity. We need to have, and to put into effect in living our lives, a deep understanding of what a community *is* and of how the life of a political community is best organized and directed if the common life of the virtues is to be made the community's mutually understood overarching goal, and sustained as such through the appropriate political institutions and practices. We need that knowledge (a practical, not "contemplative" one) in order to actually *know*, for ourselves, how to engage properly in our own individual virtuous actions, whether ones performed as political leaders or in our private affairs and private lives. This whole community, as I have emphasized, includes lots of merely decent people, and of course *they* do not have or need this full understanding of politics or of the political orientation of their decently led lives: but they too must intuitively see their lives and their decent actions as part of an interlocking, mutually supporting communal undertaking aimed at virtuous living as the highest good. That, any rate, is what I shall now argue. As I go along I will try to explain some of the intricacies of this view, which, if I am right, is Aristotle's fully developed account of the human virtues and human happiness.

In sum, then, I suggest that Aristotle has two connected but separable reasons for holding that even if we consider only his readers'/hearers' self-improvement, the project begun in the *Nicomachean Ethics* is not completed until they have studied thoroughly also the principles of legislation and political constitutions. Each of these reasons presupposes a single (as it were) body of knowledge as constituting both political knowledge and practical wisdom (as Aristotle explains in *EN* 6.8). First, he wants his hearers to know what they need to know in order to be fully effective political leaders – and thereby to live the virtuous political life, the second happiest life according to *EN* 10.8. Thus, he has in mind here a first use for that single knowledge – a use connected to what I have referred to earlier as the second of the two potential objectives for the study of ethics that Aristotle distinguishes in *EN* 1.2. But secondly, and indeed more fundamentally, he wants his hearers to know what they need to know in order to live virtuous and happy lives *as private citizens*, outside the political limelight – including, for those capable of it, the very happiest life of all, that of a private citizen who makes the activity of philosophical contemplation and theory the highest achieved goal in living their life. Here we find a connection to the first of Aristotle's objectives mentioned in *EN* 1.2. The essential point is this: anyone who expects to be good and live virtuously at all needs to understand the principles of politics, and their application, because virtuous actions and activities, however much undertaken always by individuals, are *essentially* communal undertakings.

WHAT IS AN ARISTOTELIAN *KOINŌNIA* ("COMMUNITY")?

One key but not well-understood point in Aristotle's *Politics* concerns his understanding of what a *koinōnia* (conventionally translated as "community") is. He begins the *Politics* by saying that because the *polis*[20] is a *koinōnia*, in fact the one that contains within itself and regulates all the others, it aims at some good (1252a1–7).[21] The adjective from which this noun is formed, *koinon*, means "common," in the sense of some common possession. A *koinōnia* therefore is something shared by a group of people,

[20] I leave the Greek word for "city" untranslated or adopt it into English, because of the special features of an ancient city, which occupy so significant a place in Aristotle's work: its sovereignty and wide territorial bounds.

[21] In what follows I discuss only the three subordinate *koinōnai* that Aristotle himself discusses explicitly (those of master and slave, parents and children, and the village), in addition to that of the *polis* itself. But Aristotle clearly holds that all of what in contemporary political philosophy are discussed as voluntary "associations" of different groups of citizens in clubs, religions, professional organizations, and the like, are similarly subordinate to the *polis koinōnia* and "regulated" by it in the ways I go on to explain for these other *koinōniai*. It is clear, I think, that Aristotle's conception of how common activities constitute *koinōniai*, and what it means for these activities to be common, applies equally to these associations. I leave this important application of Aristotle's theory to one side, however. It involves complications that, though interesting in themselves, would take us too far away from the more fundamental matters I am concerned with in this paper. (I have benefited from discussion with Gabriel Richardson Lear on these matters.)

as something that is theirs in common – not as a pooled sum of separate parts produced or maintained privately by each. It is a whole belonging in common as a whole to the whole group. What, however, is this thing – this *koinōnia* – that is common in that sense to a group of people? What is it, in the case of a *polis*, and in the case of the other *koinōniai* of which Aristotle speaks here, that constitutes it as a *koinōnia*?

The answer is implied by the reason he gives in this passage for saying that all *koinōniai* do aim at some good.[22] He says that this is because "everyone does *all* their actions for the sake of what they take to be some good" (1252a2–3).[23] Taken strictly, as I propose we should take it, this clearly implies that, for Aristotle, a *koinōnia*, at bottom, is some actions (in fact, some activities): all actions or activities, he is saying, including the ones that constitute *koinōniai*, aim at some good. Indeed, a *koinōnia* is some activities that the individual people making up the group engage in in common, in the way I just explained. These are activities of theirs as individuals but not with each acting on their own merely in some coordinated way so as to produce some "common" product.[24] Rather, these activities are theirs *as* group members. They are activities of the whole group (in some way that needs explanation, which I provide below), at the same time that they are, more specifically, the immediate activity on each occasion of some one person, or some smaller group of members, acting in some way defined or regulated by the specific sort of *koinōnia* in question.

On Aristotle's understanding, then, a *koinōnia* is some set of group activities performed regularly and on a continuing basis by the individual members of some group – thus sustaining the *koinōnia* in continuous existence. Of course, this is, as I said, only what fundamentally constitutes a *koinōnia* as such. Other things than activities will belong to any *koinōnia* as something common to the group, as well. In the case of a *polis*, this will include land and buildings and the contents of the public treasury, and

[22] Strictly speaking, what Aristotle says is only that every *koinōnia* "stands constituted" (*sunestēkuian*) for the sake of some good. When he adds in support of this that all of everyone's actions aim at what each takes to be some good he could be pointing to the actions of people in setting up such entities, and not (also) to actions of theirs of which the *koinōnia* itself consists. However, even so, the entity itself must be (thought by those who set it up to be) some good for themselves as human beings. And this good must, for Aristotle, consist in or include some activities, since on Aristotle's own views about the nature of the human good, activities of virtue are its core and *sine qua non*. Thus, for him, the good that any *koinōnia* is for, and of which it consists, is some activities, ones that are engaged in in common by the relevant group, which constitute it as the *koinōnia* in question.

[23] In this and my other translations from the *Politics* I follow C. D. C. Reeve's excellent translation (1998), but with many (usually unmarked) departures from it.

[24] This is clearly implied by what Aristotle says of the *polis* at *Pol.* 3.6.1278b15–17; it is or involves the shared or common life: *tēn koinōnian tēs zōēs*.

also institutions and offices structuring its political life. These make up what the Greeks referred to as *to koinon* of (that is, what is common to) the people of the given city.[25] However, these belong to the *koinōnia* because of the ways they facilitate or help to structure the group activities that are its fundamental and defining elements. They are parts of the *koinōnia* through their connection to those activities. As for the specific *koinōniai* (the city, the household, etc.), the character of the *koinōnia* itself – how it is constituted, what it is for – determines which are the common activities, and what is shared (and in precisely what ways) by the members of the group. Aristotle holds, as we see from the opening lines of the *Politics* just referred to, that the *polis koinōnia* contains within itself subordinate *koinōniai*, and in some sense controls and gives direction to them. Principal among these in theoretical importance are first, those of the household (the *koinōnia* of father, mother, and children,[26] and that of master and slave), and then that of the village, which contains the household *koinōniai* of its members, as it itself is contained in the *polis koinōnia* to which it belongs.[27] Our ultimate interest is in the *polis*, since it is there that, according to Aristotle, the activities of virtue as something to be regarded communally find their place. But given that, for Aristotle, a *polis* contains within itself the joint activities making up household and village, it will be useful, in considering which the group activities are that the *polis* consists of, and in what way it does consist of them, to begin by considering these subordinate *koinōniai* and their constituent activities.

[25] See Liddell, Scott and Jones 1940, s.v. *koinos* II 2 for the very common use of *to koinon* as a noun meaning the state or government of a given community, and (in the plural) its public affairs or public money.

[26] At first (1252a27) Aristotle mentions only husband and wife as participants in this *koinōnia* (for the first use of this term in application to the husband–wife union, see 1252b10), but later he adds the children that, as he says from the beginning (see 1252a28), are the purpose of their union: 1.3.1253b6–7, and 1.12 *passim*.

[27] Aristotle speaks at 1252a6 in terms of containment (*periechein*) of lesser by greater *koinōniai*, but in subsequent chapters in terms of composition of the greater out of the lesser (see 1252b9–10, 15–16, 27–8), and of the lesser ones as "the parts" of the greater (1253b1, 1260b13). Already at the end of chapter 1 he speaks of a *polis* as a "compound" (*suntheton*) having parts out of which it is composed (1252a18–23), and these turn out to be these subordinate *koinōniai* of household and village. It is important to realize that composition here does not mean that the whole (*polis* or village) just consists of the parts, or even *consists* of them at all. As we will see, the joint activities of which a village *koinōnia* or a *polis koinōnia* actually does consist go beyond those of the *koinōniai* that are their parts, and, indeed, the activities making up these parts become altered and extended by their inclusion in the whole. On Aristotle's analysis, the ultimate parts of any *polis* are the *koinōniai* of master and slave and husband and wife making up its constituent households. (In another sense of "part" it is the individual human beings who are the ultimate parts, see 1.3.1253a18–29: they are the sole members of or participants in the *polis*, that is, they are the ones that jointly *do* the actions of which it consists. See also 3.1.1274b38–41.)

MASTER–SLAVE, FAMILY, AND VILLAGE *KOINŌNIAI*

Before beginning (as Aristotle does) with the *koinōnia* of master and slave, I should point out that in discussing it I am following Aristotle in describing it as it is "according to nature": as it is when properly constituted and conducted, on both sides. Many actual master–slave relationships are perversions, and to those, naturally enough, what is said here about this *koinōnia* will not fully apply. It could even be that all the master–slave *koinōniai* that have actually existed anywhere were perversions. It is important in order not to misunderstand Aristotle unfairly that he should be understood as approving slavery only as it is according to nature. In discussing the other *koinōniai*, of the family, the village and the *polis*, I should be understood similarly to be discussing these *koinōniai* as they are according to nature. It is to be taken for granted that most and even possibly all the actually existing such communities have been in greater or lesser degree perversions of this natural ideal. Nonetheless, Aristotle reasonably thinks, we learn something that can and ought to regulate our own ambitions as well as our basic self-conception as we approach our daily lives, in the defective communities in which we presumably all live, if we grasp and apply what these sorts of *koinōniai* are like when they exist and function according to nature (i.e. according to the nature of human beings, and the nature of the human good).[28]

A slave for Aristotle is simply a laborer who, being stunted by birth, is capable of only a narrow range of human activities. Not only that: in doing those activities slaves (but not other adult people) require some more fully endowed human being to give them direction and keep them focused on what they are doing.[29] Aristotle says that slaves are *living* tools, and

[28] When Aristotle says that every *polis* "exists by nature" at 1.2.1252b30 (see similarly 1253a2, 25, and 7.8.1328a21–2) it is certainly not necessary, and presumably quite wrong, to interpret him as claiming that among the "natures" that are to be found at work in the natural world is the "nature" of a *polis* – "as if a *polis* is a natural entity like an animal" or a human being, as David Keyt supposes (Keyt 1991, 118–40). Aristotle can naturally, and in any event best, be taken to mean only that *poleis* come into being and exist because they are needed as fulfillments of these two other natures, which *are*, in his view, among the natures making up the natural world: human nature, and the nature of the human good.

[29] Aristotle's view is that slaves share in reason, the fundamental distinguishing mark of human nature, only to the extent of understanding what is said to them, but not so far as themselves to use anything they do understand in planning and leading their life (*Pol.* 1.6.1254b22–3). Thus they lack the power of deliberation (1.13.1260a12). This need not mean that for Aristotle slaves cannot figure out how to get *anything* done using their own thought and planning (for example, in carrying out specific assigned tasks). It may, and presumably does, only mean that anything extremely complex or requiring concentrated attention over any significant period of time (organizing a whole life, deliberating about any serious matter) is beyond their natural capacities. Their minds are always apt to wander off in pursuit of more immediate gratification.

tools for *action*: namely, certain activities, primarily, *of their masters*.[30] Slave activities, for example sweeping the floor or plowing a field or preparing a meal, are, then, on Aristotle's analysis activities engaged in in common by the individual slaves *and* the master who directs them. *We* would think the slaves are the primary agents, if not in fact the only ones in these tasks, but Aristotle thinks the master is in fact the primary agent, because it is he who directs and (ultimately) is putting his *mind* to the tasks: actions, properly speaking, require to be done by beings possessed of, and using, reason. So, on Aristotle's view, the master sweeps the floor, and so on, using the slave as his living, self-moving tool. Both master and slave are active whenever the slave works as a slave, and the actions making up the slave's work are common activities of the two. As Aristotle conceives of them, then, these activities have two agents; they are done by two people in each case: a slave and the master. And it is those joint activities that constitute the master–slave *koinōnia*. Thus, that *koinōnia* extends precisely, and only, so far as those activities do. It includes only that much of the activities that go to constitute the lives, respectively, of the master and the slave. The rest of both the slave's and the master's life are conducted outside this (very limited) *koinōnia*. Notably, in this case the good aimed at in the *koinōnia* is entirely the good of the master (and, derivatively, that of his family). The good of the slave is not at all aimed at, though incidentally the slave achieves important components of his or her good in doing their part in these common activities – to the extent that, being a stunted human being, they are capable of achieving a personal good at all.[31] All the slave's work

[30] See 1.4, which concludes with the summary statement that a slave is a human being that is a piece of property (i.e. a possession for use in actions, 1254a2) that is a tool and separate (from the human being whose tool it is – a human being's hand is a tool for his or her actions that is not separate). See esp. 1254a1–8, contrasting slave-tools (tools that *do* actions) with physical tools such as shuttles: just as it is the weaver who does some weaving *with* a shuttle (while the shuttle weaves only in an extended or secondary sense), or, to choose a different contrast, it is the person using his fingernails who does the scratching, so the master uses the slave-tool to sweep the floor or cook the meals, or dig the trenches for a barn's foundations, etc. In all these cases the agent, or primary agent, in the actions is the one that uses the tool, not the tool, even in the case of the slave, who, being a human being, is also an agent active in the doing of the action. One should compare so-called "master-craftsmen" (*architechtones*) in relation to under-craftsmen or assistants (*hupēretai*) who do the actual labor of the craft under the hands-off direction of the masters (see 1253b38–1254a1). On Aristotle's view "even in the case of actions involving external objects [such as weaving some cloth or sweeping a floor] the one who does them, in the fullest sense, is the master craftsman who directs them by means of his thought" (7.3.1325b21–3, trans. Reeve with one change); Aristotle expresses the same view more compactly at 1.13.1260a18, where Reeve seems to misunderstand the grammar and reverses subject and predicate: "the work that is done is in the first instance that of the master craftsman."

[31] See 1254b17–20 (slaves are "people whose work (*ergon*) is to use their bodies," this being "the best thing to come from them"), and 1252a30–4 ("the same thing is beneficial for both master and slave"); and 1278b30–7 (rule by a master is "rule exercised for the sake of the master's own benefit, and only coincidentally for that of the slave").

is aimed at making the daily lives of the master and his family go well, both by providing the materials and the material conditions needed by the family to sustain their lives, and by assisting them in engaging in some of the activities that make their own lives up, but in which slaves do not themselves engage jointly with them.

The common activities constituting the *koinōnia* of husband, wife, and children are importantly different. Psychologically they go much deeper. In addition to being done by more than one agent, as with the master–slave activities, all of these are aimed at a good common to all the participants. In fact, this good includes goods common to the participants at two levels or in two different ways, as I will explain below. The result is that the common activities of family members living together in a household are activities done by them in common in a much deeper way than the activities common to slaves with their masters.

Aristotle says that female and male form a couple for the sake of procreation.[32] They do so not by deliberate choice (not that in most cases there would not have been one) but out of the desire arising naturally in human beings as in other animals to leave behind offspring like themselves. What then are the common activities of the resulting family members that make up such a *koinōnia*? The common activities of the couple, as a couple, will include their sexual activities as marriage partners, and all the activities of raising and educating the children, even if those are performed primarily by only one of the parents at any time (in some instances with the use of slaves). Raising and educating their children is a common project,

[32] 1252a27–30. He says that this coupling (*sunduazesthai*), like others, such as that of master and slave, brings together people "who cannot exist (*einai*) without one another" (1252a26–7). He can hardly mean either that individual women or men could not, once raised, exist without being members of a male–female couple, much less that the male or female sex in general cannot exist without each male or female forming a lasting couple with some member of the other sex. Perhaps he has in mind, reasonably enough, simply that *some* males in each human generation must form couples with *some* females (not all the males and females can live apart by themselves) if either males or females are to continue to exist. Still, one might wonder why, if that is what he meant, he thinks that this continued existence should require the sort of enduring couples he is speaking of here, and not mere one-off sexual couplings. The latter would seem to suffice. In view, then, of what he goes on to say about the household and the male–female *koinōnia* at its center, it seems better to suppose he has in mind that human beings need to be *raised* (nurtured, educated, initiated into a whole culture) over many, many years in order to become full adult male or female human beings at all (and not, maybe, crude and wild unhuman animals looking physically like human beings). Since (he thinks) this requires being brought up within a family with a male head and his female partner, there cannot be male and female adults in any subsequent generation if there are no such families for them to belong to as children and grow up in. It does seem plausible that in that sense human males and females cannot (continue to) exist without adult males and females living with one another in households (even if our modern societies have developed ways that make possible, in special circumstances, the raising of children ready for a full adult life as human beings even outside such a rigidly defined family).

undertaken by the parents together. When the mother, say, is helping a young child to learn to play fairly and with due consideration of the other children he is playing with, it is an essential component of what she is doing, implicit though not normally self-conscious, that this is part of a whole series and set of activities that fit together to constitute a larger and more extended activity she and the father are engaged in together over many years, of raising the child to adulthood. Some of the components of this single activity are performed in the first instance not by her but by the father; and all of these, whichever the primary agent may be, are endorsed and actively supported by both parents.

Other activities, too, are included in the husband–wife–children *koinōnia* of the family: all the activities of daily life together within the household, the meals taken together, the conversations, the games played, and, of course, with particular emphasis, those of these into which the moral virtues (as Aristotle understands them) are integrated, since those are the center and substance, for him, of a well-lived human life. (I return to this moral component in a moment.) It would not be easy to specify more exactly which the activities are that constitute this *koinōnia*.[33] Still, they clearly make up only a relatively small part of the activities of its individual members that make up the totality of the daily life within the household. Each of the family members spends most of their time in activities of a private and personal sort that though taking place within the household are not done as common activities joined to the activities of the other family members in the way I have indicated.[34] Hence, the *koinōnia* of the family, like that of master and slave with which it is joined in the household, is one of severely limited scope.

[33] For Aristotle, the constituent activities of households (and those of the two *koinōniai* that make them up) are included in the activities of villages, and both of these are also parts of the *polis* and its activities. This complicates any attempt to specify which actually *are* the constituent activities of the family *koinōnia*. Aristotle offers a genetic account of cities as coming to be through a progressive natural development and articulation of human life, in households and villages, in ultimate pursuit of the final and complete human good. It is necessary to abstract the family from these encompassing other institutions and their activities if we wish to consider the family *koinōnia* on its own. So considered, the common activities of family members will be limited to the overtly shared ones of eating, talking, playing, etc., together.

[34] It may be worth noting here that when we consider family life as taking place within a complete *polis* context, where both it and the *polis* life of which it is part are conducted fully according to nature, we will think of all the daily activities of the family members within the household, even these private ones (the parents sitting in their rooms reading, or whatever, for example), as belonging to the family community's common activities, as parts of the common activity of the *polis* as a whole. While reading in their rooms, the parents conceive of what they are doing as part of an overall shared life. Where, as explained below, the shared life of virtue (or decency) is pursued as everyone's highest good, even such private activities take on a wider dimension of community.

All the activities that do form part of the family *koinōnia*, however, aim at the good of all the participants, and at a good held in common by them all. As noted above, this marks a significant departure from the common activities of master and slave. There are two dimensions to this community of good. First, the activities themselves are good, because well conceived and well carried out (remember: we are discussing a family according to nature). This good, the good of the activities themselves, as such, belongs to and is achieved by both or all of the participants simply in doing them. It is furthermore an indissolubly common good, consisting not (or not only) of a pooled sum of individual goods achieved separately in the actions of the separate agents. It is a single good belonging in equal measure to each of the participants, because it is a good achieved by the pair or group of participants acting together. In this sense, the mealtimes, and the conversations, are taken up with an activity that the family all engage in together, as a common undertaking – well conceived and well carried out by the members in their own individual ways, each doing something different which fits together with and is (at least implicitly) understood by them all to be a contribution to some activity in which they are engaging in common, and of which each individual contribution is a constituent part. The good therein achieved is a good common to them all. The good of or in the well-conducted conversation or meal taken together, for example, is a single good accruing equally to them all.

Many of these activities are also, of course, aimed at goods external to the activities themselves. This is a second level or way that there are common goods aimed at in the activities. The meals are aimed, among other things, at obtaining daily sustenance. Relaxation and stimulation are further external goods provided at mealtime, as well as through conversations at other times, and games and other pastimes. These external goods may, and mostly will, be distributed individually to the individual members and will not be something indissolubly common, as the good of the common activities, as such, is. In this case, to say that the good is a "common" good means only that the provision of these respective separate shares *to all* is part of what the activity consists in, what it is for. The relaxation and stimulation of games and daily social interactions, as well as the sustenance provided at meals, are for all the members of the group – but one by one. There are, however, some such external goods that are aimed at as common goods in a stronger sense, instead. Most notably, the parents' activities in raising the children aim (as an external objective) at making them good human beings and enabling them to live good human lives as adults. But one's children being good and living well, as good human beings, when adult is part of the good of any parent.[35] So in this case even the external good is something that belongs, when it is achieved at all, to the parents in common, not in a divided way – just as with the common good achieved in the doing of the common activities.

When we turn to the village *koinōnia* we find that Aristotle says extremely little about it. Almost all he says is that whereas the household is "naturally constituted for the everyday," the first *koinōnia* (in the analysis of a *polis* from the simpler to the more complex) "constituted for the sake of other than everyday needs" is the village, itself constituted out of some number of households.[36] I take the reference here to "other than everyday needs" to mean the following. Villages make possible a social life, with a wider and more interesting range for conversation and other leisure-time interaction than single households do. By introducing local cults with priesthoods, and festivals (with poetry readings and drama performances), and the like, they also expand the range of human activities. These new human activities belong specifically to village *koinōniai* and are not possible within a separated household. They satisfy other than everyday needs – other than needs to do with sustenance, reproduction, basic security, and bodily comforts, including minimally necessary clothing and implements for maintaining human life with basic needs satisfied. With the reciprocal exchange of surplus production, however, that villages introduce, villages also make possible satisfying these everyday needs more easily than life in an isolated family could, and more satisfactorily, too, because of the resulting greater variety and higher quality of materials and material goods they make available for consumption and use.

We should also take account of the fact that, for Aristotle, village *koinōniai* are made up of household ones, as I mentioned. This means that (as I have just implied) the common activities constituting the household receive a wider context that makes them involve the pursuit of a wider common good than just that of a single family. Parents are raising children to live well not just in their own households but in the villages of which their households are parts, just as the householder is directing his agricultural slaves and his farm animals for sustaining the life not only of his own household, but in part also (reciprocally) that of the other households in the village. Thus the life activities definitory of the household will, in general, also become, in this expanded form, activities of the village *koinōnia* as well, since the life of the family and its household can now be

[35] See *EN* 8.12, "parents feel affection for their children as being something of themselves" (1161b18), they "love their children as being themselves, for the ones coming from them are, as it were, other selves of theirs" (b27–9).

[36] See 1252b12–13, 15–16.

seen as part of the life of the whole village. These activities, focused in the first instance on the common good of some single family's members, are implicitly conceived of also as part of a wider common enterprise engaged in together with the members of other village families: one aimed, both in terms of everyday and other needs of life, at living well *as* a whole village. Thus not only the religious and cultural and expanded economic activities I referred to, but also the more local ones of everyday household life, can be seen as engaged in in social union with all the neighbors making up the other households of one's village.

THE *KOINŌNIA* OF THE *POLIS*: THE MORAL LIFE AS COMMUNAL PRODUCT

Finally, we reach the *koinōnia* of the *polis*. Aristotle says that the *polis* "has reached the limit of total self-sufficiency" for human life. It makes possible, and itself actively supports, a life for its citizens in which human nature becomes fully developed and human capacities for action are completely fulfilled. Beyond this he says very little – less in fact, surprisingly, than he does about the much simpler *koinōniai* of the household. So in this case, the crucial one for us, we will have to think through for ourselves, without significant textual clues, in what the communal character of this *koinōnia* and its activities consists.

To his remark about the self-sufficiency of the *polis* Aristotle famously adds: "It comes to be for the sake of living, but it remains in existence for the sake of living well."[37] "For the sake of living well" (*tou eu zēn charin*) here means not for the sake of living at a high economic standard, or even that plus the provision of a rich array of cultural expressions to appreciate and participate in. All such considerations belong to (mere) life at a high level, not to "living well," as Aristotle intends that expression here. The "living well" aimed at in a *polis* means, specifically, having its citizens live a life that is governed by their possessing the human virtues, each to some significant degree, though presumably more fully in some citizens than in others (I say more about this below).[38] They aim together through their lives in the *polis* at decency and virtue as their highest good, their fullest natural self-realization. The citizens of a *polis* that is "according to nature" all place the highest value on virtue and virtuous action, and they possess the virtues

[37] 1252b27–30.

[38] See 1.9.1257b40–1258a14, where Aristotle contrasts two ways of understanding the correct and natural goal of the craft of money-making: obtaining and preserving enough for "living well," i.e. as he says explicitly there, virtuously, or making as much money as possible, with a view simply to "living" – i.e. luxuriantly and indulgently.

of habituated feeling and action (and that of practical understanding too) as nearly fully, individual by individual, as is realistically to be hoped for in any *polis*-sized human population. They are all, at a minimum, *decent* people, according to Aristotle's account of decency in *EN* 10.9, as explained above.[39] This means that the citizens structure their lives through exercising their virtues, as a matter of individual independent judgment, on a constant and regular basis in all that they do.

Thus, according to Aristotle the *polis*, as an institution, comes into being through the union of a number of villages in some self-contained territory that possesses a city center, external trade relations, and large-scale cultural and religious institutions. Hence the (so to speak) mere life (as opposed to the life of virtue) that the *polis* makes possible is far richer and more interesting, more completely fulfilling of human natural capacities, than that of an isolated village – in just the ways that I said above that village life is richer than an isolated household's could be. This enrichment consists not only in new aspects of life belonging explicitly to the level of the city and carried out in the city center (most notably overtly political institutions and activities), but also in an expansion and enriched content of the activities definitory of household and village *koinōniai* that result from their being fit into this new context. It is the needs felt by people

[39] As noted above, at *EN* 10.9.1179b18–20 Aristotle speaks of those who are decent as having a "share" in (*metalambanein*) virtue (*aretē*). These, he explains, are people who have firmly established good habits of feeling and action, and who love the fine but do not have any well-developed practical knowledge of the human good, such as his lectures on ethics and politics provide for those able to follow and make his arguments truly their own. Thus Aristotle recognizes merely decent people as in some way or at some level possessed of virtue (*aretē*). In the *Politics*, at 7.1.1323b23, he again speaks of virtue – he means specifically moral or ethical virtue, virtue of feeling and action – and practical wisdom, and happiness as well, as coming in degrees. Taking my cue from this, in speaking of the "virtue" of the citizens of a *polis* according to nature in what follows, and of their "activities of virtue," I speak loosely or comprehensively (as Aristotle himself does in this context). I include within the scope of these activities not only the full virtue of those who possess all the Aristotelian virtues of character as well as of practical intellect (and even, in some persons, the virtues of theoretical study and contemplation), but also the lesser virtues of habit and feeling attained by decent ordinary citizens living nothing but ordinary daily lives with ordinary political participation, but no leadership whether intellectual or political. Aristotle is plainly aware (see *Pol.* 7.7) that human populations vary in their inborn and natural capacities for developing virtues (see 1327b33–6). This must be taken into account in any description of an Aristotelian *polis* according to nature. Nothing further can be theoretically required of all the citizens of such a city than that they all be decent people. Given the "share" in virtue that decent people have, it is legitimate for Aristotle to speak, as he does in this connection, of the "virtues" and "virtuous activities" of all the citizens, these decent ones as well as the few fully virtuous. Presumably, when he says in *EN* 2.1.1103b18–26, that though we human beings are not born with virtues of feeling and action we are by nature able to receive them (through habituation), it is only the virtues of decency that he means to indicate that we all are by nature capable of receiving. The other virtues (and even full and true virtue of feeling and action) may well be beyond the natural powers of some people. A *polis* according to nature must be conceived on that basis.

living in households and villages for the richest possible such mere life that, as Aristotle implies, explains the coming into being of cities. They come to be for the sake of the higher standard of living and richer cultural and social life they enable. But when led now in the context of a *polis* life, the mere life of villagers comes to include, as was the case with the village, completely new activities as well, most notably, as just remarked, the activities of the shared self-government of the citizens, through specifically political institutions and activities. Included also are all the wider social, religious, and cultural activities, plus the wider range of work opportunities and interesting personal relationships that come from the foreign trade and larger-scale economic activities the *polis* makes possible for its citizens.

The enrichments I have mentioned so far concern only the "mere" life that a *polis* makes possible. But, as we have seen, the *polis koinōnia* once formed *exists* for the sake of a life lived by the individual citizens that is governed by their human virtues. These are, in the first instance, the practical virtues, those of character and practical intellect – at whatever level and in whatever degree they individually may possess them (see n. 39). They will all be at least decent people. Some will possess these practical virtues to the fullest – including the virtue of practical wisdom, based, as we saw above, on practical knowledge of all that Aristotle tries to bring the readers and hearers of his *Ethics* and *Politics* to know about the human good. And, among those, some will be the political leaders, living the "second happiest" life of *EN* 10.8 (see above, pp. 225–7). Yet others will even lead the happiest life without qualification: the life of full practical virtue, but devoted to philosophical knowledge and theory as their actually achieved highest good.[40] These varied levels of virtue and types of happy

[40] At least, this seems to be an implication of Aristotle's discussion in *Pol.* 7.1–3 of the "most choiceworthy life" and of whether it is the same for an individual person and for a *polis* community (1323a19–21). Aristotle does not there openly state this result, perhaps out of a sense of delicacy in addressing his audience, where those aiming at political careers presumably predominate. But he clearly indicates there, as he explains more fully in *EN* 10.7–8, that his own view is that the most choiceworthy life for an individual is the contemplative one of theoretical, especially philosophical, study (see 1325b14–30); and he also leaves no doubt that whatever life is most choiceworthy for an individual is also most choiceworthy for cities (1324a7–8, 1325b30–2). That means that his account of how the people of a *polis* that is completely self-sufficient for human life will live must include the provision that among them will be a group of citizens who live the contemplative life (and so are provided an education that will enable them to live that way). By living that life individually and as a group, they contribute directly to the life of the *polis* of which they are a part, so that (provided other conditions in it are met as well) it will live the most choiceworthy life for a *polis*. That he leaves this upshot as something for the alert reader to see does not make room for doubt about his intentions: the city according to nature, and more especially the city that "fulfills our prayers" of book 7, will provide a place for contemplative philosophers among its other political goals in seeking the happiness of its citizens. (For this distinction, see below, n. 53.)

life among the people of a *polis* according to nature will become important below, but for now we can set them aside and focus simply upon what it means that all the citizens of this "natural" *polis* are virtuous people (at a level at least of decency), and together succeed in living well and thus achieve that for which *poleis* exist.

It is crucially important to notice that when Aristotle says the *polis koinōnia* is for the sake of living well, he is conceiving of living well (i.e. virtuously) as the central common activity of the citizens of a *polis* (that is, one that is constituted and functions according to nature). As we have seen, for him, a *koinōnia* simply is at bottom a set of common activities. When he declares that a *polis* is for the sake of living well he makes virtuous activity the central common activity defining the *polis koinōnia*. Living virtuously corresponds for the *polis* to a family's shared daily activities in the household and the other than daily ones shared in the *koinōnia* of the village – the local religious and cultural and social and economic activities of the village. But in this case, as part of its total self-sufficiency for human life, *all* the actions and activities of the *polis* members are included in the resulting *koinōnia*, since virtue itself (at any rate practical virtue of character) provides the motivation that is expressed in and causes the doing of *all* of its possessors' actions. The life of virtue that the *polis* aims at and makes possible is thus a person's total life (insofar as that consists of actions and activities they freely engage in). In this it differs from the lesser *koinōniai* of household and village. In the latter, selected common activities take place within a larger context that includes many private activities of each participant that are not, or need not be, shared with others. This consequence follows from Aristotle's conception of the practical virtues as governing the whole of one's life: it governs all one's preferences, the relative evaluations of all other goods besides the good of virtue itself, and in consequence the basis for all one's decisions and actions. Thus, the shared life of virtue that Aristotle maintains the *polis* is for includes all of their shared household and village activities, now conducted throughout in accordance with the virtues of character and practical intellect – or, in the case of the merely decent citizens among them, at least with good habits of non-rational feeling, including an active love of the fine, and some rational motivation (*boulēsis*) inclining them to do the actions required by the practical virtues. But it includes many new activities too, including explicitly political ones – and much more, as well. It is all-encompassing.

Just like the activities making up the household and village *koinōniai*, then, Aristotle conceives of these activities of the virtues, central to the life of a *polis*, as common ones, engaged in by the citizens in common.

The activities of virtue in the *polis*, though most of them will be the work of some single individual, constitute a common enterprise pursued, in the first instance, as with village and household activities, for the sake of the good inherent in those very activities (in this case, the activities of exercising the virtues). These are to be conceived, somehow, as a common good for all the participants, and not a divided one, of which each citizen would get only a private share, separate for each.[41] In the *polis* as it is according to nature the citizens conceive of themselves as each pursuing (and, indeed achieving) his or her own personal good (the highest good of living constantly in the exercise of the virtues in all their individual actions, choices, practical judgments and attitudes), but only through pursuing that good as a part of the common pursuit, along with all the other citizens, of the virtuous life of the *polis* itself, that is, the virtuous life of all the citizens. Aristotle is conceiving this common good, of which the good realized by each in their own virtuous actions is a part, as achieved by all of them acting together. Since it is achieved by all of them acting together (somehow), the whole of it – the good of the city, namely the citizens – becomes a single good achieved by them all. So much is implied by the idea, which I have developed above in discussing the common activities making up the lesser *koinōniai* contained within the *polis* of how activities constitute *koinōniai*. The citizens' activities of virtue are the common ones constituting this overarching *koinōnia*, the one that is sufficient for the complete development of the capacities belonging to human nature, and so for the human good, and a happy life, for all – at whatever level and in whatever degree they possess the virtues that make a happy life possible. But how are we to understand that the virtue and virtuous activity of *all* are in fact to be pursued by *each* as common goods – goods in which all share? How does one do one's virtuous actions and live one's virtuous life as part of some common good shared by oneself and all one's fellow citizens? Are we to understand that (as with the common activities of master and slave) in some way each virtuous action of any individual citizen is itself an accomplishment, not of that single person alone, but of the whole moral community of the *polis*? Is each citizen somehow a co-agent with each other one in the doing of their virtuous actions? Or is it simply (as with the common good pursued in the activities of child-rearing and daily conversation and other interactions in the family) that each does their own actions in such a way as to interlock with those of others and thereby

[41] As with the household and village activities, many of these will also be aimed at achieving external goods for those participating in them, but these will often be common only in the sense that each is intended to get their own private share in each case. See my discussion above, pp. 237–8.

constitute a larger good, that of the whole community's living virtuously? In short, when Aristotle maintains that we should see the lives of virtue led by the citizens of a *polis* that is constituted according to nature, as consisting of virtuous actions undertaken somehow in common, as *shared* actions of them all, how are we to understand this? There are two connected questions here. What does it mean for all to be involved in the doing of the actions of each? What does it mean for each, as a result, to share in the good achieved by his fellow citizens, one by one, in the good achieved in *their* actions?

We can begin to answer these questions by noting that each of the citizens, in participating in this *koinōnia*, thinks of his or her own scheme of values, contained in the virtuous outlook on life with its assignments of relative and comparative value to all the goods available to a human being, including virtuous activity itself, as something he or she shares with their fellow citizens. This scheme of values, they think, is not just something each has come to understand as correct through one's own personal experience and education – as a matter of one's private moral insight. They and their fellow citizens have made a common and mutually agreed decision to support this scheme of values. It is something they all, individually and collectively, have come to *understand* (to some degree: fully, by some of them, less than completely by others) as the correct one for human beings to live by. It forms the basis of their city's legal system, and of their agreed and common conception of what is just and unjust in the designing and implementation of institutions of self-government. It governs their system of public education and their criminal law and its administration. In living according to their scheme of values each conceives of that scheme as a *common*, mutually agreed and understood, basis for all of them, together, to be living. It is on that basis, and with that implicitly in mind, that each does all their particular acts of virtue.

Moreover, one's fellow citizens do not just espouse as correct one's own scheme of values but can be seen themselves actually to live according to it. So, with the increased strength of commitment that comes from seeing one's values actively affirmed by one's neighbors, in a common way of life, one can be assured (as assured as any human being could ever be) of the truth of one's own moral beliefs. Other human beings not only say they see things the same way but show they believe it by the way they live. That widespread agreement in practice is strong evidence that these beliefs, and this scheme of values, does derive from a correct use of reason itself and is not some merely arbitrary social invention or some other aberration. With that assurance, they can count on themselves, as they might well not otherwise be able to do (on this see further below), to carry out unwaveringly their commitment to acting always virtuously, whatever the difficulties or pressures of circumstances might be. They see their own views as not something private to themselves and a few other people like them, or something merely a matter of how "we" in a certain family or of a certain class live and pride ourselves on living. These are a whole *polis*-sized population's shared reflective judgment about human life. By manifesting in their own actions and steady way of life their common moral convictions, therefore, each of them lends support to each of the others when that person's own efforts in the common enterprise require some significant personal cost, or loss. Each of them, implicitly at least, sees their own virtuous actions, especially when undertaken under significant pressure of circumstances, as part of a mutually supportive system for sustaining the similar actions of all the others. They recognize the moral benefits to themselves along with the others in thus seeing their own virtuous actions as undertaken in this spirit of a common and shared way of life.

In this way, they can think of their own, *and* the others', acceptance of and commitment to, and their ability to sustain, a life of moral virtue as something to which *all* of them contribute equally. In each virtuous action done by any of them there is a strong psychological background of support coming from the regular behavior of all the others in which they evince their shared commitment to this way of life and the scheme of values it rests upon. Each virtuous action performed by any of them has part of its own causation in the regular virtuous actions of the others. Thus, each one is right to think that the good they achieve for themselves in their own virtuous actions is also the product of a joint effort together with the others, aimed at the common good for all of a virtuous life led by them all.[42] There is (most often) a single agent when any virtuous action is

[42] Constance Meinwald (in discussion) proposed an analogy with string-quartet playing to clarify how I conceive of the joint character of the individual virtuous actions of each participant in this common enterprise. It is possible, she pointed out, for quartet-players to produce the common music by each playing in time with the others, as a sum of the four separate parts, but without "playing *with*" one another: to play with one another is for each to be listening to each of the others, adapting their individual playing as they go along to that of the others, so as to achieve a musical product that is common in a stronger way – a mutually conceived and throughout mutually adjusted, single shared conception of the musical product being aimed at. The household activity of raising children, as I presented it above, offers an application: it is not merely that the mother and the father agree as to who will be responsible for which activities of child-rearing, conceived of as a common enterprise.

done, but that agent in so acting is drawing upon support coming from the common and visibly shared commitment of all to this way of life and to its supreme value for a human being. In that way, both the act itself and the good that it achieves (in being an act of virtue) is shared in by all the others as well. The act and the good are legitimately seen, and felt, by the agent, and by everyone else who knows of it, as a common accomplishment. As a result, the good each achieves in his own virtuous actions is not limited to the good in those actions themselves but expands to include shares in the further moral good that it helps to bring about in the actions of all the others.

Education in this system of values and in living according to it begins, of course, in the home, in the daily life of the household, as children are being raised by their parents to be morally well-functioning adults. But, as Aristotle once says,[43] children and others in a household must be educated "with an eye to the constitution." Whatever virtues they acquire in their household life, and learn to exercise there, must be calibrated to the larger life they will lead as members of the political community of which the household is the smallest and, in one way, basic part. This is one important aspect of what Aristotle means by saying at the outset of the *Politics* that the *polis koinōnia* contains within itself and controls or regulates the other *koinōniai*. The household has as one of its necessary functions to provide a

Each pays attention, if not constantly then at least regularly, to what the other is doing and makes adjustments, as they see them needed, to their own contributions. There is a constant interplay of discussion and mutual adjustment in the ongoing shaping of the common enterprise, even though for the most part the constituent activities are undertaken by one single parent. The case of the activities of moral virtue differs, however, in that morally virtuous action is always a second-order superposition upon some first-order activity. In some cases the first-order activity may itself involve Meinwald's sort of "playing with" – as in playing a competitive game. Here, if it is played with due concern for the quality of the play itself and not just for winning, and so with the exercise of relevant moral virtues, we find "playing with" at two levels. In other cases, say when one eats a solitary meal but does so with the virtue of temperance fully engaged, there is "playing with" only at the second-order level. The agent coordinates their activity implicitly with other virtuous actions of their own and others at the same and other times. But there is no "playing with" at the first-order level. Hence, the application of the analogy to the case of moral action involves special complexities.

[43] *Pol.* 1.13.1260b14–17. At *Pol.* 5.9.1310a12–18 Aristotle says that of all the ways one might attempt to make a constitution survive the most important is for the citizens to be "educated in a way that suits their constitutions" – differently in oligarchies, which place an exaggerated value on money and wealth, from in democracies, which value equality and espouse fairness conceived of as equality of status, wealth, access to political rule, etc. In the correct constitutions, and especially in the sort of "polity" that, as I understand him, he expounds in book 7 as the best under conditions one could pray for, the education in the home and in the public schools will be conducted not just with an emphasis on virtues, according to a *true* conception of which states of character actually are virtuous, but with a correct conception of the highest human good as simply *being* a life of virtuous activity.

proper context for children to receive training in the virtues. But this must not be seen as directed merely toward the proper regulation of the goods of daily life with one's intimates in an extended family. That life is indeed the essential province of the household *koinōnia*. All the educational activities of the household must be carried out as activities taking place, no doubt, within the household – but as belonging to the life of the specifically political community of the *polis* itself, not life in a household wrenched for purposes of theory out of that wider context. They are an education for conceiving of the decent way of life the children are being habituated to want to live, both inside and outside the confines of the household, as a communal undertaking, in which each person in all their own decent behavior is also giving support to, and reciprocally receiving support from, the decent behavior of others.

This, then, is how I propose we should understand Aristotle's admittedly brief and quite unelaborated account of the *polis* as a *koinōnia*. We can see now how Aristotle's theory of the *polis koinōnia* proposes for its citizens a common good larger than the good each could have achieved simply by living a socially and psychologically isolated life of virtue. Each participates, as a sort of co-agent, in the good consisting of all the other citizens' virtuous activities, as well, of course, more directly and intimately, as agent of their own. They participate in this larger good, and can claim a share of it, not, or not merely, because they act upon a mutually shared system of values, in which each does their part in helping to achieve the good belonging to some common project, in the way that participants in an interesting conversation or other cooperative activity, such as a well-played game, do. They mutually give each other significant support, drawn on by each as they do their individual virtuous actions, through the commitment to those values that each exhibits in their turn through their fine actions, and through the motivation for and understanding of the virtuous life as the human good that they display in doing them. This support does not make each a full co-agent in the virtuous actions of others, as with the master in relation to the slave's work. But it does mean that each is implicated in all the actions of all the others, as an approving and supportive partner, much as in the interactions of a mother and father in their cooperative activities of child-rearing, or in a good conversation (despite the differences, noted above in n. 42, between moral activities and the results of these other cooperations). And the common good it makes possible is not, like the goods of these other cooperative activities, some encompassing new activity, or set of activities, to which the different activities of each of the participating agents contribute and together constitute; it is the very same good, namely the good of virtuous activity, but now a good belonging to others than the single primary agent him- or herself.[44]

POLITICAL KNOWLEDGE/UNDERSTANDING AS NECESSARY FOR SELF-IMPROVEMENT

Let us now return to Aristotle's argument in *EN* 10.9. There he argues that the approved hearers of his lectures on ethics need to continue their studies through hearing lectures on politics. As we saw, he clearly implies that they need that further study, in major part, in order to become fully good and virtuous people, and so, to be able to live completely fulfilled and happy lives. This is the first reason he gives. But we saw a second reason, too, one not clearly stated as such, but apparent to us on reflection: they also need it because they (or most of them) are expecting, and he is expecting them, to become political leaders in their communities. In that role (and, possibly, as adviser to other communities, or to monarchs or other dynasts), they will have to have and apply this further knowledge if they are to succeed, as they will wish to, in improving the ordinary citizens' lives, whether of their own or other cities. We have now seen that on Aristotle's account of the moral or practical virtues, as it unfolds in the *Politics*, living the life of virtue in a *polis koinōnia* (its natural home) is in fact a joint project, in which all one's individual virtuous actions are understood, and conceived as one does them, as parts of a larger common project, with one's fellow citizens, in living virtuously as a whole community. The *polis* is the self-sufficient *koinōnia*, whose *raison d'être* is to make possible (and actual), if it is organized and lived according to nature, the fully good human life. There, it is a life of virtue lived in common, in the way I have sketched.

We need now to examine the bearing of this conception of virtuous living as a communal undertaking on the first of these two reasons that Aristotle gives in *EN* 10.9 why those studying ethics with him need to complete their studies through investigating with him the first principles of politics. Aristotle says this is needed if one is actually to *live* virtuously. How are we to understand that claim? We can begin by setting aside as irrelevant the possibility that, owing to a combination of lucky circumstances, someone might acquire the virtues through being brought up in

[44] This needs qualification as regards the intellectually virtuous activities of the contemplatives of the given *polis*. These are not activities of just the same sort (i.e. practically virtuous) as the ones the ordinary citizens, or even the political leaders, contribute to this common good, the good that gets thereby shared in by all the fellow citizens. See below, pp. 260–3.

an excellent family and, owing to unusual native insight and strength of character, be able as adults to live a full life of virtue even in an isolated location, or in a thoroughly corrupt polity where there exists no sufficient body of like-minded others with whom to form the kind of community Aristotle presupposes in his account of a *polis* according to nature. Aristotle's claim should be taken to relate not to what might or might not happen if someone is extraordinarily lucky. Rather, it concerns what happens in the nature of things, given human nature (even when perfected through natural processes – but not lucky ones), and the natural world at large (including what can naturally be expected to hold among a normally diverse human population). Our question then is: Why does Aristotle think anyone who is going to live a life of virtue (practical virtue, in at least the first instance) needs to grasp the first principles of politics, and their application to the social and political organization of life in a polis?

In addressing this question we need to bear in mind that, as I indicated in my account of the common life of virtue led by all the citizens together of a *polis* according to nature, there are significant differences of level and extent of virtue characterizing different ones among the citizens.[45] They all share the same conception of the human good as consisting in virtuous living, and they all live virtuously and so happily – but some at the level of mere decency, others at that of full virtue; among the latter, some are political leaders, others not; and among these *others*, some lead private lives devoted to the pursuit and practice of philosophical knowledge, at its deepest, while the rest lead more normal private lives devoted to family, friends, socially productive work, and cultural activities of a less rarefied kind, and the like. These differences will affect how one answers this question in relation to the different sorts of person and life.

We can start by considering the political leaders.[46] These, as I have said, live the secondarily happiest "political life" of *EN* 10.8. I have mentioned

[45] Aristotle's repeated insistence that a *polis* is necessarily and essentially a union among people of different kinds (see *Pol.* 2.2.1261a22–30 and 4.4.1290b37–1291b13) reflects this. It is easy to see that the members of the different groups he lists in the 4.4 passage will vary quite significantly in level and extent of virtue, even under the best of naturally achievable circumstances.

[46] Interestingly, in his description of the city of *Pol.* 7 and its institutions and constitutional practices, Aristotle makes no special provision for any such group. It is apparently not part of the formal constitution, any more than it was at Athens. Officially, all the citizens rule and are ruled in turn (and not merely in the contrived sense that Aristotle explains in 7.9.1329a2–17 and 14.1332b32–41 – ruled when young, ruling when old). But that is, of course, compatible with there being a group of people who seek and exercise leadership in ruling year by year. They take the lead in the assembly's affairs and offer themselves, when others might not, for the most important offices. In fact, that everyone rules and is ruled in turn is not a very demanding practice, important though it is from the point of view of justice. There is a plethora of offices to be filled year by year in any Greek *polis*. Since Aristotle holds that this life of political leadership is the undiminishedly happy one of practical

just above that these people need to know political science (*politikē epistēmē*) in order to do their chosen work properly, that of leading their cities well (as well as offering advice to those outside who need it). It is true, also, that they have to have and exercise in their leaders' activities the full practical virtues, of character and intellect. The knowledge they need for the task of political leadership is not some merely theoretical grasp of political science. It has to be a practical knowledge; indeed, as we have seen, Aristotle holds it is nothing other than the same knowledge that constitutes practical wisdom, and in having practical wisdom, as Aristotle argues, they will also have to have, and will have, the full virtues of character too. And possessing these virtues leads them to *want* to take a leadership role – not only because of the good that, being virtuous, they want to do their fellow citizens, but more especially because, as this knowledge makes them realize, the exercise of their virtues as political leaders provides the widest and deepest scope for exercising them. As Aristotle says, the good of a city is "a greater and more final or end-like thing" and "finer and more godlike" to achieve than one's own happiness alone (*EN* 1.2.1094b8–10). And these are consummate lovers of the fine.

So far, however, this sounds rather formalistic. Yes, Aristotle is committed to all these doctrines, about political knowledge and practical wisdom having the same content, and being practical not theoretical, and about the mutual presupposition between practical wisdom and virtue of character. And these do commit him to holding that in order to live virtuously and happily the citizens capable of and suited for political leadership will learn the principles of politics, if they can, and use them both in doing their political work and (in the form of practical wisdom) in leading their whole lives (not just in doing that work). And Aristotle's claim that citizens of the *polis* according to nature (including these leaders) live their life of virtue as a common project in shared virtuous activity implies that the leaders will intend all their virtuous activities, both the leadership ones and the rest, as their own contribution to the common life, and the common good, brought about by all, in part by these virtuous actions of which they alone are the primary agents. But what is it about virtue and virtuous living that requires of them that they see things that way? Nothing in their role as political leaders, bent on making the life of their city as happy as possible, and so as virtuous as possible, and therein aiming to attain their

virtue, his city according to the political expert's prayers (the one of book 7) must be understood as providing room for such a group living that life. The consequence of not providing it would be that the city could not live, as Aristotle clearly thinks it does, the most choiceworthy life for human beings.

own happiness, seems to require that they do that. They might care deeply about virtue, their own and that of others (virtue, to be sure, in the latter case, mostly merely at the level of decency), but as essentially two separate concerns. They could conceive of virtue as a private possession by each one who has it, opening up to him or her the good of virtuous activity, where that good is a personal and private accomplishment of each. They might exercise their own virtues on that basis, in the expanded and deep way that only using them in activities of political leadership makes possible, without conceiving of themselves as joined with the rest of their fellow citizens in a common project of sustaining a common and shared life of moral virtue for all, and a common good provided to them all through that means. In their virtuous activities of leadership they could be aiming at the good of their fellow citizens entirely as an external objective of their actions, while at the same time aiming, as a separate, internal, or constituent end, at their own good, simply *in* acting virtuously, both in those leadership activities and in all the others making up their life.

What reason do they see, then, for going the one step further demanded by Aristotle's theory of the virtuous life as a common one lived together with one's fellow citizens? Here again one must speculate a bit. Two things come to mind. First is something not so much a matter simply of the nature of virtue itself, or of normal human nature as what needs to be perfected by virtue, but rather a significant and interesting consequence for human beings of regarding virtue in the communal way. Even though these leaders can have the satisfaction and achieve through their actions the fineness of seeing their city flourish in moral terms under their leadership without making their actions part of a *common* life of virtue in union with their fellow citizens, that leaves the good of and in the other citizens' lives an external good for them, something they bring about but do not participate in. They do not get that good as something personal to them, in the way that the people performing the virtuous actions themselves do. It is something they can take credit for, it is a good for them, but still it is the same rank of good as the other good achievements they attain through the use of their virtues in other aspects of their lives. (It is a good at the second level, as described above, pp. 237–8.) But when they pursue their life of virtue as part of a common enterprise, sustained externally through their efforts as leaders in directing the educational and other institutions of their city – an enterprise engaged in also in their own ways by the others, including ones less politically and morally well accomplished than they – they come, as I have explained, to share directly and internally, and not in that mere external, productive way, in the good consisting in all those other virtuous activities. That expands and deepens the range of good that they directly participate in, by expanding the virtuous activity in which their own virtue gets expressed. Their virtue now expresses itself also *in* the virtuous activities of the others with whom they live this common life of virtue. As noted already, they exercise their virtues in this communal way not only in their private lives, but also in their actions in directing the city's life. This combination gives them the widest possible scope for virtuous living. This doubly direct, personal, participation in *all* the good that the virtuous activity of the whole *polis* constitutes, brings with it a final, complete, absolutely fully perfect, life of virtue – in fact, as Aristotle says in *EN* 10.8, the *happiest* life possible of practical virtue.

I think that is one powerfully effective consideration these leaders will see, giving rational support to the life of shared virtuous living to which they will in any event have been brought up. Furthermore, this life would be, and have to be, grounded in the full knowledge of political science. This is obviously so for their leadership activities (it would be required even if they did not regard those as parts of a system of shared activities in pursuit of a communal end). Less obviously, and apparently less extensively, this same knowledge would ground all their other actions, as private persons and in their private lives. This is something, as we have seen, that Aristotle commits himself to when he adopts the position that political science and practical wisdom are the same body of knowledge. Practical wisdom provides, and reasonably so, on Aristotle's theory, the fundamentally necessary grounding for action that is fully good, fully in conformance with the requirements for virtue of character. But now we can see why Aristotle may be right to insist on that identity. If all the leaders' virtuous actions – including their daily lives as private persons, with their families and friends – are to be conceived of (in part) as their particular contributions to a common activity of living virtuously, undertaken and supported by all the citizens, they *need* at least a deep knowledge of the basic principles of political science in order to carry them out properly. (This is something, it would seem, they would *not* need if they did not regard their virtues and their everyday virtuous activity in this communal way, but only as their personal possessions and goods.) That knowledge contains the full account of what it means that human beings are *polis*-living creatures ("political animals"), and of how a *koinōnia* is constituted – how the different groups that a *polis* needs to have in it relate to one another in their contributions to and shares of the good consisting in this common enterprise. This they clearly need, in order to perform virtuous actions in this communal way.

Required also, at least in general terms, is knowledge about the varieties of political constitution and their merits and demerits, and about the one that would be best under optimal, at least in principle naturally available, circumstances.[47] Of course, these latter, more or less "technical," details are not needed in any specific way in many or most of their private activities. Nonetheless it is reasonable of Aristotle to hold that this whole, complete, body of knowledge, extending to all the technical details, must pervade their every virtuous action, providing a constant framework of understanding in which each action of virtue would be seen as falling into its proper place. If this complete knowledge of human nature and the human good were lacking in an agent – whether a political leader or not – as the intellectual grounding of their actions, they themselves and those actions could not claim the full perfection of human nature that, properly conceived, virtue expresses. And Aristotle's reasonable demand is that the political leaders of a *polis* existing according to nature should be fully virtuous persons.

That, then, is a first consideration helping us to see the political leaders' reasons in taking the step of conceiving of their own virtuous lives as parts of a common project of living virtuously as a whole community. There is a second consideration as well, this time a matter simply of the nature of virtue itself and of normal human nature as what needs to be perfected by virtue. This applies not only to the political leaders but to all the other citizens as well: the fully virtuous ones who do not choose lives focused on government and public service, including as a subset the fully virtuous ones who live lives of philosophical inquiry and knowledge, plus the rest (the bulk) of the population, who are at best fully decent people committed to a life of virtue at the level of decency. It has to do with the essential psychological fragility, belonging to human nature as such, of any human being's commitment to the moral life.

As we can see from what he says in *EN* 10.9, as discussed above,[48] Aristotle thinks that most human beings, however well developed in the virtues, retain the human tendency to act for immediate pleasure or to

[47] I have in mind here primarily the lengthy and varied questions that Aristotle takes up in books 4–6 of the *Politics*, concerning oligarchy, democracy, and tyranny, as well as aristocracy, "polity," and kingship. But there are many details concerning the nature and functions of citizenship (judicial, deliberative, administrative) taken up in book 3, as well, the knowledge of which seems quite remote from such matters as how to eat one's food or carry on a civil conversation with a shopkeeper or hired employee, or deal fairly with others in the myriad circumstances of daily life.

[48] Here one needs to bear in mind (see above, pp. 220–5) the distinction Aristotle draws at *EN* 10.9.1179b7–20 among three classes of people: (1) the "many" who are permanently only ever going to behave decently through legal requirements, backed by pleasurable incentives and painful sanctions; (2) the people who, having been habituated well, can come to acquire "some share of virtue" through argument, because of their love of the fine, and so become decent people living decently from their own inner resources, without the need for constant appeal to such incentives and sanctions; (3) the ones who can become truly and fully good, in important part because they have a native love of the fine that is strong enough to permit argument to make virtue take full possession of their souls. (See also 1180a10–18.)

avoid short-term discomfort – the tendency that training in virtue seeks to overcome. He proposes publicly promulgated laws as one remedy. This liability is found particularly among the young, but also among Aristotle's "many" (the unregenerate mass of human beings). But as he explains there (1180a1–4) he thinks it applies even to one who has acquired both good habits and sufficient practical knowledge and understanding of the human good to be living a committed decent life. Hence, even decent people, if left entirely to guidance by their private judgment, would inevitably sometimes lapse. They would fall away into unvirtuous choices and acts. However, their personal and private virtues can be expanded so as to become part of the psychological basis for a communal life devoted to virtuous activities by all, as a shared undertaking, as I have described it above. In that case, they would experience the support from the community that I mentioned (pp. 244–5 above) – deriving not just from the abstract knowledge that others agree with one in supporting the same basic system of values, but from the shared and felt participation in a common commitment to those values in an interlocking common pursuit of them in all one's own, and their, virtuous actions. This would give them the added psychological boost they would need in order to more nearly overcome this apparently permanent tendency of human beings to yield to the attractions of immediate pleasure, even when it is not decent to do so. When facing such attractions they would feel not only their own rational wishes against indulging, but also the force of the whole community's decision, in which they participate, against doing so. They would also wish to continue supporting the resolve of others through their own action's resolve. They would fall away into unvirtuous actions less often and less disastrously. Hence, a life of virtue led in that communal way would be a less defective way of living virtuously than if one led it thinking of oneself only as a single person aiming at one's own single happiness through living virtuously. It would be a psychologically more secure one, and therefore also one with fewer lapses from virtuous action. Hence, Aristotle seems to suggest, this communal life of virtue would bring with it for the political leaders too individual lives that were more completely virtuous than they could otherwise achieve.

Now, to be sure, fully virtuous people, such as the political leaders, are not nearly so much in need of psychological support through seeing their attitudes about virtue shared by others as these more ordinary, less completely virtuous persons. It seems possible that Aristotle thinks that his authorized hearers might even reach a point in their ethical-political development where they were no longer subject to any attractions of immediate pleasure in circumstances where acting as it would incline them to would lead them to depart from virtuous action. This could happen either because they would not find any pleasure in acting that way then or because their inner psychology is sufficient to give them reliably other motivations strong enough that they never do depart from virtuous action, even if they did sometimes feel it. Certainly, that is the ideal view Aristotle takes in describing fully virtuous persons. On the other hand, at *Pol.* 3.16.1287a30–2 Aristotle says that "appetite is like a beast, and spirit (*thumos*) corrupts rulers, even if they are the best men" – which is why "law [rule by which is better than by absolute rulers] is understanding without desire," that is, understanding without the possible corruption due to appetite or spirit which individual persons are subject to, and which haunts rule by them. That even the best men in ruling are subject to corruption (and so, to give way to vicious action in some circumstances) suggests that even the political leaders of the best city will remain actively vulnerable to being carried away by appetite or spirit into acting unvirtuously.[49] The ideal of the *Ethics* may be *too* ideal: actual, normal human nature, even when perfected, may not allow single individuals, considered singly rather than as members of a community, to achieve this immunity or near-immunity from moral contraventions.

Now, presumably, Aristotle thinks that this reveals a feature of full moral virtue in itself, independently of whether the virtuous person in question is a political leader. He is saying that, for human nature as it is even at its most perfected, in actual human beings in expectable and normal circumstances of life, we find this moral fragility. His point would be that not only the leaders but all fully virtuous persons in the *polis* – all the fully virtuous people in the city according to nature, including those leading lives of philosophy, and the others, as well, who do not opt for the political life – are subject to this psychological frailty. If so, then the moral support given to decent people, by living a fundamentally communal decent life, would be needed and would have its salutary effects also in giving any and all fully virtuous persons reasons for making their own life of virtue part of a shared undertaking of all the citizens for virtuous living. They too could not consistently and constantly engage in virtuous activity without that moral support, and so without joining their lives with those of the others

[49] Notice that this does not mean that Aristotle's allegedly virtuous people would on his own analysis turn out to be merely self-controlled. Self-controlled and uncontrolled persons are ones whose characters are weak, making them subject to regular and characteristic temptations to self-indulgence under certain recurrent circumstances (see Cooper 2009, 10–14). The lapses here in question are uncharacteristic, but do sometimes occur, in no characteristic pattern – or would occur, if people were left simply to their own devices, relying only on their own self-generated inner commitment.

(including the merely decent people) in their *polis koinōnia*, by making the virtuous life of the whole community an internal objective in their pursuit of their own good through virtuous activity (both in their private affairs and in discharging their public responsibilities). Hence, we could infer a second reason the political leaders could see for taking the further step Aristotle demands of them. They would thereby make their lives more securely and completely virtuous – and so, happier too.

So much, then for the political leaders. They need to know both the first principles and all the details of political science, as Aristotle investigates those in the *Politics*, not only because they need them in giving direction to the educational, social, and political activities of their city, with a view to helping others to live as virtuously and happily as they individually are capable of living. They need them also in order to be virtuous (fully virtuous, as Aristotle says at *EN* 10.9.1179b2–4) themselves and lead the happiest lives that they, individually, are capable of living – the political life in which the highest good they achieve is found in their virtuous activities of political leadership. Furthermore, they need them even in order to have and exercise – consistently, reliably, and with full and unremitting commitment – the virtues in the rest of their lives, as private persons, as well. It seems true that they will not actually *use* much of the detailed knowledge of political regimes, their varieties, strengths, and weaknesses, and suchlike, in this part of their lives. Nonetheless, they certainly will use – constantly – the basic principles of politics and the essential political aspects of the human good in living their lives as virtuous persons. In fact, simply *qua* virtuous, they will have to possess a fully expanded knowledge of these political matters, since having that knowledge is one of the human perfections in which being (fully) virtuous consists. Even if the detailed knowledge is not being applied directly in many situations when they act virtuously, the whole body of knowledge does stand behind and gain expression in each of their virtuous actions, whether in their role as political leaders or not.

As we have seen, however, Aristotle does not say in *EN* 10.9 merely that this political knowledge is needed by political leaders. It is true that in the *Politics* Aristotle does seem often (for example when he speaks, as he very often does, of what "the statesman" or political leader, *ho politikos*, or "the legislator" will need to take into account or do)[50] to address himself most particularly to those of his readers/hearers who are preparing themselves

[50] See *Pol.* 1.11.1259a33; 2.5.1263a39; 2.7.1266b27; 2.11.1273b6 and 11; 3.3.1276a34; 3.13.1283b37 and 1284b17; 4.1.1288b27; 4.12.1296b35; 4.14.1297b37; 5.9.1309b35; 6.5.1319b33; 7.2.1324b24 and 1325a8; 7.7.1327b38; 7.14.1332b35, 1333a14, 35, and b35; 7.16.1334b29, 1335a6, 1335b14; 7.17.1336b5; 8.1.1337a11.

for political lives.[51] But he means to make the same claim about anyone who is to acquire the full moral virtues and live a life successfully guided by the recognition of virtue as the highest human good. In terms of the citizenship of Aristotle's city according to nature, these will include as many others, besides the political leaders, as possess the full virtues of character and practical intellect, and, most particularly, the contemplative philosophers among them who (ideally anyhow) live Aristotle's absolutely happiest life. It obviously does not include the large body of citizens who are merely good, in the sense of committed, decent people: such people, almost by definition, do not possess any developed *knowledge* of practical matters, or political either. Being decent and living decently, on Aristotle's view, does not require *knowledge* at all. However, as Aristotle says, decent people do "have a share in" virtue (10.9.1179b19–20), and this suggests that if anyone who succeeds in being fully virtuous and living fully virtuously must possess and use political science in living their lives, then something parallel must hold for decent people. Aristotle must think that they need some unsystematic and perhaps only intuitive, but real, grasp of at least the basic principles of political science in order even to be committed, decent people. And from what I have said already about the life of virtue conceived of as a common project of *all* the citizens of a *polis* organized and lived according to nature, one can see why he might think this. That unsystematic grasp of the principles of politics, where politics is conceived of in Aristotle's way, is something one might reasonably think needed for decent people to do their part in contributing to this common life, through conceiving their own decent lives as parts of a larger common enterprise in morally good living by all.

In fact, of course, as Aristotle himself clearly recognizes,[52] the citizens of any city will be, morally speaking, a very mixed bunch. This applies also to a *polis* that exists according to nature, an idealization in which all the citizens will be virtuous people at one level or degree or other, living virtuous lives of one sort or another.[53] Hence we should treat these

[51] It is possible, even likely, that our *Politics* text was originally written as an investigation and exposition of political theory, specifically just for persons interested in that subject on its own – including prominently anyone who had in mind to take an active role in politics as administrator or political adviser. When Aristotle wrote the *Nicomachean Ethics*, however, with its explicit orientation of ethical theory toward and as part of political theory, understood somewhat more widely, he invited us to read his *Politics*, enlisted now in this new project, in a wider perspective – as addressed not only to intending political leaders but others as well, concerned to be as good persons and live as happily as possible.

[52] See *Pol.* 7.1.1323b23, cited above, n. 39.

[53] A city "according to nature," sketched in book 1, is to be distinguished from one that "fulfills our prayers" (i.e. *tēn kat' euchēn ginomenēn*, 4.11.1225a28–9), that is, the prayers of experts in political

categories – the merely decent, the fully virtuous who are political leaders, the fully virtuous who devote themselves to philosophical study and knowledge, the fully virtuous who live neither political nor contemplative lives – only as establishing benchmarks for theoretical evaluation and judgment. We need to bear this clearly in mind in considering how Aristotle may think his claim that political knowledge is needed if one is to become good applies to others than the political leaders. We need to think of the "decent" people, or again the "fully virtuous" ones of different categories, as specifying a complex range of types of person in each case.

The "merely decent" must be taken to include people who may have done little or nothing more in their moral self-formation in adulthood than reflect on and accept their (in fact) more or less correct moral upbringing, so that they come to "wish" regularly for and decide on virtuous actions, on the basis of an established, though entirely vague and quite unargued, intellectual sense that virtue is the highest human good[54] – but to include also many who have gone deeper in their thinking, perhaps by studying moral philosophy, and in any event with the result that they have a more articulate and intellectually robust grasp on some of the true reasons why that does actually constitute the highest good. Even these more seriously thoughtful people count as merely decent, however, because they do not possess the full knowledge that Aristotle attempts to help his authorized readers and hearers in the *Ethics* to achieve: that is, as we have seen, the virtue of practical wisdom. On the other hand, there is no clearly marked or even easily definable line to draw between such a more seriously thoughtful decent person and the least complete true possessor of the practical knowledge that Aristotle counts as practical wisdom. We must recognize a range of accomplishment there too, in the true grasp of the philosophical principles that establish human virtue as our highest good.

science, whose knowledge will only show to the fullest what it is capable of achieving if certain external conditions of land, character of the population, location, etc. are provided to them (as Aristotle says at 7.4.1325b37–1236a5): these come by luck (7.13.1332a29–32, the rest of what is good about the resulting city comes from the knowledge and skill of these expert advisers). Aristotle outlines his own version of such a city in book 7. In doing so, he assumes only conditions that are naturally possible, not, as he complains of Plato in his "ideal" cities of *Republic* and *Laws*, naturally impossible ones, however much one might wish the impossibility away (7.4.1325b38–40, cf. 2.6.1265a17–18). That city is of course also one according to nature; but, as I understand him, the same holds for any other "correct" constitution (on these, see *Pol.* 3.7).

54 One could think of these people as equivalent to the ones John McDowell mistakenly claims are in fact possessed of practical wisdom – people with nothing more than a strong sense of certainty ("Surely you can *see* that is wrong?") that their inherited way of life is the best one. See McDowell 1979, 331–50.

Presumably even the political leaders in the *polis* as it exists according to nature will exemplify a range of moral accomplishment: though all of them are more or less fully good (not merely decent) persons, not all of them would on an acutely narrow conception count as having absolutely complete and perfect practical wisdom.

First, then, about the merely decent. It is reasonable for Aristotle to hold, as I have suggested above, that these people need *some* degree of (if not knowledge, then) intuitive grasp of at least the basic principles of politics, because that grasp is surely the minimum necessary for them to be able to join in the common activity of living according to moral virtue that constitutes the *polis koinōnia* in which they participate. The same considerations apply in their case, as with the political leaders, for developing and seeing their virtues in this communal way. The good consisting in the virtuous activities of all the other citizens (including those of the leaders and the other fully virtuous people) become thereby internal objectives of their own virtuous actions. In that way they come to participate actively and have a share in the good of the actions of those others. That is one way in which by living their virtues in this communal way, and not as a merely personal and private matter, they expand their virtues themselves and enrich and make more complete as virtuous activity the activity of virtue in their lives. As I mentioned already, they also obtain the benefits of psychological support in sustaining their own virtuous commitments commented on above. In that way the place of virtuous activity in their lives, being more deeply entrenched in their minds and characters, is made more secure and constant than it would be if it were instead only some private and personal pursuit. Their lives become more complete lives of virtue, wherever on the range of decency their lives may fall. In order to live that way they need to understand the *polis* as an overarching *koinōnia* possessing and regulating villages and households, with their different structures and functions, as constituent sub-*koinōniai*; they will need to bear in mind, as they go about their daily lives, some basic reasons why the correct form of *polis* to live in is one that treats virtuous living for all its citizens as its organizing goal. Only so will they have a sufficiently live sense of their virtuous actions as forming part of the common life of the *polis*. Beyond these very general ideas they will not need to go, of course. They do not know – even at a level of intuitive grasp – and do not need to know detailed points concerning the failings, or needs, of other types of *polis*, or anything about intricacies of policy-formation, of the sort that Aristotle investigates at length in his *Politics*.

The non-political fully virtuous fall into two relevant groups: those who live the contemplative life (for Aristotle, one higher and better than even the political life), and those who forego the political life instead for the practice of practical virtue alone, but at a level less expansive than that of the political life, with other work and other occupations than either philosophical or political ones. Let us first consider this second group, leaving the contemplatives aside for now, since their situation raises interesting additional questions.[55] About these we can be brief. It is true that, like the merely decent, they will not need political knowledge for the first reason that the political leaders will need it: for giving direction to the city's political affairs and decisions. But like the political leaders, and for the same reasons, they will need to have and even, in a background way,

[55] One might perhaps doubt that I am right in supposing that Aristotle's theories envisage such a group – or, if they do, whether they can do so consistently. If Aristotle holds, as I have inferred he does, that the happiest life devoted to the life of practical virtue (without theoretical philosophy) is the political life (see above p. 226), and if any fully virtuous person must know that it is (since each fully virtuous person possesses the intellectual virtue of practical wisdom, of which this knowledge is a part), does not that imply that *all* fully virtuous persons would at least strive to live the political life and reject any other life (aside from the contemplative, to which they cannot aspire because of lack of native talent or, perhaps, personal predilection) as unworthy of them? Is not Aristotle's *Nicomachean Ethics*, then, after all, aimed solely at an audience of intending political leaders, offering them the prospect, as well as the training needed, of this (second) happiest life of political leadership in their cities? However, if so, that might well open up an inconsistency in Aristotle's theories of its own. Presumably even in the city according to nature there would not be room for as large a number of people to live the life of active leadership as there could naturally be of fully virtuous persons living in it. The consequence would be that this excluded group of citizens would be frustrated and (relatively) unhappy. But this city is supposed to make every citizen happy (as happy as their specific natural abilities permit). In fact, the same considerations of personal predilection and natural talent that might prevent a (morally) fully virtuous person from pursuing the life of theoretical philosophy might also quite reasonably lead them not even to want to pursue a political life. After all, a successful political leader needs to be personable and outgoing and have other related talents and personal characteristics, as well as political science, even when the latter is conceived of as a kind of practical knowledge. It is people like that that I have in mind: pursuing a political life, they can see, does not suit them and might leave them somewhat unhappy, while a non-political life would be quite satisfying. One should bear in mind two points. First, Aristotle's announcement of the political life as (second) happiest comes only at the very end of the *Nicomachean Ethics*. No one reading the long discussion of happiness and the virtues in the preceding books would think the sort of person discussed there was being conceived of throughout as a political leader (necessarily). Indeed, the level of discussion there involves no reference at all to (so to speak) professions or lines of work or favorite occupations; the characters defined are clearly being conceived of as encompassing lots of different sorts of life, so far as such issues are concerned: *fully* virtuous lives of lots of such different sorts are clearly being envisaged. (If not, Aristotle could not be thinking that his lectures could have the wide appeal that they do, and that he surely knew and intended them to have.) Secondly, his account in the *Politics* of the life of moral virtue as a shared one already widens considerably, even for fully virtuous people not leading a political life, the scope of the engagement of their virtues in their life in just the way that, viewed from the *Ethics*, the political life might seem uniquely to do. The "gap" between the political life and the fully virtuous one focused on other sorts of work or occupation, when seen in the light of the theory of community in the *Politics*, is not so very great.

make use of, a full political knowledge (including all the detailed points) even in their private and daily lives (see above, p. 256).

Let us turn finally to the contemplative philosophers. What function does political science have for them in leading their lives? What need do they have of this knowledge, in order to be fully good and live fully good and happy lives? One of Aristotle's firm doctrines in ethical theory is that one who achieves the virtues belonging to and perfecting theoretical thinking must also have acquired (so to speak, first) the practical ones (see *EN* 10.8.1178b5–6; 6.13.1145a6–9). This is not the place to inquire into his reasons for thinking so, or to consider fully the plausibility of this view. However, we should bear in mind the following points. To pursue and achieve the good that exists in the activity of theoretical understanding is not merely to possess and exercise some information grounded in some theoretical reasons one grasps. It is to exercise the virtues of and for theorizing *in* acquiring and activating that understanding. The good here in question consists in the activity of those virtues in that understanding. Aristotle holds that acquiring and exercising those virtues presupposes the possession (and, in their proper sphere, the regular and constant exercise) by the person in question of practical wisdom and the (full) ethical virtues. We might, ourselves, think of the acquisition and use of the two sets of virtues as entirely independent and unrelated, even if we could agree with Aristotle in thinking of the full perfection of a whole human person as including both. One reason why he disagrees with us and insists on the close interconnection of the two, with practical wisdom seen as enabling and supporting the virtues of theory, is the following.

A person's practical wisdom does not, of course, *direct* the activities of contemplation when he or she engages in them: *sophia* or theoretical wisdom does that. It is through *sophia* that one experiences the good of those activities. But practical wisdom (not *sophia*) is the virtue on the basis of which one knows the value of that activity, in relation to the values of other goods, both good activities and their good products or accomplishments, and good relationships with others and good experiences undergone. This good – the good of "theorizing" – is one among the first-order goods that the second-order good of practically virtuous activity oversees and organizes into a properly virtuous overall life. Even when engaging in excellent contemplative activity, if that good, on that occasion, is to form part of a well-lived life, one must also be exercising one's practical understanding of the way that what one is doing is good, and how that goodness compares with and relates to the other goods of human life. Lacking that (though it might seem a bit paradoxical to say this) one would not in fact achieve in one's activity the good itself that (otherwise) it would contain. It would be flawed as an activity of contemplative thought however, so to speak, technically competent it might be. And since it belongs to practical wisdom to know about the perfection of the *whole* human person, anyone who possesses that virtue will necessarily *not* think that obtaining the good of theorizing at just any and every moment when it might be available to us (despite its being the very best *single* good) would make for a well-lived life. The cost in neglect of the legitimate social, bodily, and other goods, of concern to any whole human person, however lesser they may be taken one by one, would be prohibitive.

In any event, it is clear that Aristotle does hold that the contemplative philosophers in his city are practically virtuous people, as well as theoretically so. Hence, they need the knowledge of political science in just the same ways as I explained three paragraphs back; I will not repeat them here. From the present point of view, the contemplative philosophers live in just the same way as the other fully virtuous non-political leaders of a city. Since they all live their moral lives on the communal basis I have outlined, they will experience their practical virtues as having the wider effects that go along with seeing them as parts of a common basis for living, shared also with the other citizens. Their practice of virtuous actions is aimed not just at their personal good in acting virtuously but at the good of virtuous action also when done by others sharing this common objective and this common practice. And as we have seen, Aristotle reasonably thinks that that way of common living is impossible without the possession and use of the full knowledge of political science. It is true, of course, that, unlike the political leaders, the contemplatives do not use their virtue and their practical knowledge in directing the life of the whole community through its political institutions. Hence they will not act within that wider and finer context for the exercise of their virtues, which the political leaders enjoy. However, they are much more than equally compensated through the consummate goodness of their theoretically excellent activities of contemplative thought and study, which the political leaders correspondingly miss altogether. They also receive, to the small extent that it may be needed, the support in their practice of the virtues that comes from seeing others joined with them in the active affirmation of the supreme human value of virtue.

There is one final point to notice about the lives of the contemplative philosophers, unique to them. I said above that in their case their practical wisdom oversees not just their morally virtuous activities, as with the rest of the citizens, but also their activities of theoretically excellent thinking and understanding. Not only is it the function of their practical wisdom to know about excellent theoretical understanding *that* it is the highest good for a human being (that holds equally for any other fully good person: both the political leaders and the other non-political fully good citizens discussed above). In their case, practical wisdom is exercised constantly in each particular activity *of* theoretically excellent thinking and understanding: such activities are in each case, in different ways, the products of both of the two sorts of virtues of thought, practical and theoretical. Since all the citizens agree with one another in the common pursuit, together, of the virtuous and happy lives of the whole community, the good these philosophers achieve in their activities of contemplation takes its place, alongside the good of their morally virtuous activity, as a contribution to the good aimed at in common with all their fellow citizens. Hence, through the cooperative and common virtuous activities of their fellow citizens that support the philosophers in their philosophical endeavors, especially the activities of political engagement aimed in part at sustaining the presence in the *polis* of philosophical inquiry, this good becomes a common good for all the citizens. The other citizens do not engage in philosophical inquiry themselves, but through their virtuous political support for its presence among them, they come to share in the good that it contributes to the common life led by the *polis* itself, that is, by all the citizens together. In this way, the life of the *polis*, that is, the life lived in common by its citizens, becomes one in which the fundamental human goods inherent in that life, the goods of virtuous activity – excellent contemplative thought and study, as well as morally or practically virtuous activity – are, to some degree, shared by them all. It is not merely a common life, but a common life in which all participate in all the fundamental goods that human nature makes available to human beings.

In this final section I have undertaken to understand Aristotle's reasons for his claim in *EN* 10.9 that no one can become fully good without possessing and using the full knowledge of political science, including the knowledge of constitutions and legislation that he himself investigates in the *Politics*. I have argued that Aristotle recognizes in actual human beings, even ones who achieve the perfection of human nature that possession of the human virtues constitutes, a continuing openness to being drawn to indulge in pleasure or "spirited" satisfactions even when it is wrong to do so. We can overcome this weakness, inherent in human nature, under normal and naturally attainable circumstances, to the extent that it can be overcome at all, only by joining together with other like-minded people in a common pursuit of virtuous living as a joint project and effort.

His account in the *Politics* of the *polis* "as it exists according to nature" describes such a community. The need for political knowledge as a basis for virtuous living can be understood and even justified if we accept these rather plausible further Aristotelian ideas and theories. In presenting and elaborating this interpretation I have gone considerably beyond anything Aristotle himself says explicitly, or, in part, even clearly suggests, either in the *Nicomachean Ethics* or the *Politics*. I hope that I will at least have shown some of the moral and philosophical resources, and the considerable interest, of Aristotelian ethics and politics.

Allan Gotthelf's contributions to classical philosophy

PUBLICATIONS

1976/7: "Aristotle's Conception of Final Causality," *Review of Metaphysics* 30; repr. with "Postscript 1986" and additional notes, in (i) *Philosophical Issues in Aristotle's Biology*, ed. A. Gotthelf and J. G. Lennox, Cambridge University Press, 1987 and (ii) *Essays in Ancient Philosophy*, ed. R. Bosley and C. Panayides, Nicosia: Intercollege Academic Press, 2005.

1979/80: Review: N. P. White, *A Companion to Plato's* Republic, *Review of Metaphysics* 33.

1980/1: Two-part essay review: M. C. Nussbaum, *Aristotle's* De motu animalium, *Journal of Philosophy* 77; *Review of Metaphysics* 34.

1981: Short reviews: J. Findlay, *Plato and Platonism*; E. McLain, *The Pythagorean Plato*; W. Sahakian, *Plato*; and N. P. White, *A Companion to Plato's* Republic; in *Bibliography of Philosophy*.

1982: "Divine Comedy [humorous passages in the Aristotelian corpus]," *Ancient Philosophy* 2.

1983: "Teaching Aristotle's Ethics," *Teaching Philosophy* 6. Includes reviews of: A. O. Rorty, ed., *Essays on Aristotle's Ethics*; J. M. Cooper, *Reason and Human Good in Aristotle*; W. F. R. Hardie, *Aristotle's Ethical Theory*, 2nd edn.; F. Eterovich, *A Commentary on Aristotle's* Nicomachean Ethics.

Review: R. Sorabji, *Necessity, Cause, and Blame*, *Journal of the History of Philosophy* 21.

"Arius, Theophrastus and the *Eudemian Ethics*: Comments on Professor Fortenbaugh's Paper, with Special Attention to *pathos* in Aristotle," in *On Stoic and Peripatetic Ethics*: *The Work of Arius Didymus*, ed. W. W. Fortenbaugh, Rutgers Studies in Classical Humanities 1, New Brunswick, NJ: Transaction Books.

1984/5: Review: M. Boylan, *Method and Practice in Aristotle's Biology*, *Review of Metaphysics* 38.

1985: *Aristotle on Nature and Living Things: Philosophical and Historical Studies Presented to David M. Balme on his Seventieth Birthday* (editor), Mathesis Publications (US) and Bristol Classical Press (UK).

"Notes towards a Study of Substance and Essence in Aristotle's *Parts of Animals* ii–iv," in *Aristotle on Nature and Living Things* (see just above).

1987: *Philosophical Issues in Aristotle's Biology*, editor (with J. G. Lennox), Cambridge University Press.

"First Principles in Aristotle's *Parts of Animals*," in *Philosophical Issues in Aristotle's Biology* (see just above).

"Postscript 1986" and additional notes to reprint of "Aristotle's Conception of Final Causality" (above, 1976/7), in *Philosophical Issues in Aristotle's Biology*.

1988: "*Historiae* I: *Plantarum* et *Animalium*," in *Theophrastean Studies*, ed. W. W. Fortenbaugh and R. W. Sharples, Rutgers Studies in Classical Humanities 3, New Brunswick, NJ: Transaction Books.

1989: "The Place of the Good in Aristotle's Natural Teleology," in *Proceedings of the Boston Area Colloquium on Ancient Philosophy*, vol. IV, ed. J. J. Cleary and D. Shartin, University Press of America.

"Teleology and Spontaneous Generation: A Discussion," in *Nature, Knowledge and Virtue: Essays in memory of Joan Kung*, ed. R. Kraut and T. Penner, *Apeiron* Special Issue.

"Aristotle as Scientist: A Proper Verdict," cassettes of two-part lecture, Irvine, CA: Second Renaissance Books (available from Amazon.com).

1991: *Aristotle's* Historia animalium Books VII–X, ed. and trans. D. M. Balme, prepared for publication by Allan Gotthelf, Loeb Classical Library, Harvard University Press.

"A Note on the Loeb *Historia animalium*, vol. III," *Ancient Philosophy* 11.

1992: *Aristotle: Metaphysics, Biology, Ethics: Report of the 1988 NEH Summer Institute* (editor), New Haven: Council for Philosophical Studies.

"Report on Recent Work and an Additional Bibliography," in *Aristotle's* De partibus animalium *I and* De generatione animalium *I (with passages from II.1–3)*, trans. with notes by

D. M. Balme, Clarendon Aristotle Series, revised ed., Oxford: Clarendon Press.

1992/3: Review: A. C. Bowen, ed., *Science and Philosophy in Classical Greece*, *Review of Metaphysics* 46.

1993/4: Review: W. W. Fortenbaugh *et al.*, eds., *Theophrastus of Eresus: Sources for his Life, Writings, Thought and Influence*, 2 vols., *Review of Metaphysics* 47.

Review: Aafke M. I. Van Oppenraaij, ed., *Aristotle:* De animalibus. *Michael Scot's Arabic–Latin Translation. Part 3: Books XV–XIX:* Generation of Animals, *Biblioteca Orientalis* (Netherlands).

1996: Abstracts of: (i) "Aristotle as Scientist: A Proper Verdict (with emphasis on his biological works)," (ii) "The *editio maior* of the Greek Text by D. M. Balme", (iii) "Book Order, Chapter Divisions, and A Phantom Dedication," in J. Brams and P. Beullens, "*The Tradition of Aristotle's* De historia animalium. *Report of the International Workshop, Leuven, 30 May – 1 June 1996*," in *Bulletin de Philosophie Médiévale* 38.

1996/7: "In memoriam: A. C. Crombie," *Review of Metaphysics* 50.

1997: "Division and Explanation in Aristotle's *Parts of Animals*," in *Beiträge zur antiken Philosophie: Festschrift für Wolfgang Kullmann*, ed. H.-C. Günther and A. Rengakos, Stuttgart: Steiner Verlag.

"The Elephant's Nose: Further Reflections on the Axiomatic Structure of Biological Explanation in Aristotle," *Aristotelische Biologie. Intentionen, Methoden, Ergebnisse*, ed. W. Kullmann und S. Föllinger, Stuttgart: Steiner Verlag.

"Understanding Aristotle's Teleology," in *Final Causality in Nature and Human Affairs*, ed. R. F. Hassing, Washington: Catholic University Press.

1998: "Love and Philosophy: Aristotelian vs. Platonic," cassette recording of lecture, Irvine, CA: Second Renaissance Books (available from Amazon.com).

1999: "From Aristotle to Darwin: Closing Words," in *Aristotle's Animals in the Middle Ages and the Renaissance: Proceedings of the 1997 Leuven Conference*, ed. C. Steel and G. Guldentops, Leuven: Brill.

"A Biological Provenance," contribution to a memorial symposium on Montgomery Furth's *Substance, Form and Psyche: An Aristotelean Metaphysics*, *Philosophical Studies* 38.

"Darwin on Aristotle," *Journal of the History of Biology* 32.

2002: *Aristotle's* Historia animalium, vol. 1: *Text and Critical Apparatus Books I–X*, ed. D. M. Balme, prepared for publication by Allan Gotthelf, Cambridge Classical Texts and Commentaries Series, Cambridge University Press.

2005: "David M. Balme," *The Dictionary of British Classicists, 1500–1950*, ed. R. B. Todd, Thoemmes Press.

"William Ogle," *Dictionary of British Classicists, 1500–1950*, ed. R. B. Todd, Thoemmes Press.

Reprint: "Aristotle's Conception of Final Causality," in *Essays in Ancient Philosophy*, ed. R. Bosley and C. Panayides (above, 1976/7).

2007: "Theodorus Gaza's Translation of Aristotle's *De animalibus*: Composition, Influence and Date" (with Pieter Beullens), *Greek, Roman and Byzantine Studies* 47.

2008: Review: C. C. W. Taylor, *Aristotle's* Nicomachean Ethics *Books II–IV, Translated with an Introduction and Commentary*, *Notre Dame Philosophical Review*.

2010: Comments on Leunissen, "Aristotle's Syllogistic Model of Knowledge and the Biological Sciences: Demonstrating Natural Processes," *Apeiron* 43.2–3.

Mariska Leunissen and Allan Gotthelf, "What's Teleology Got to Do with It?: A Reinterpretation of Aristotle's *Generation of Animals*," *Phronesis* 55.4.

CONFERENCE, INSTITUTE, AND WORKSHOP ORGANIZATION

1981: Co-organizer (with M. Boylan), Marquette University Conference on Aristotle's Philosophy of Biology.

1983: Co-organizer (with D. M. Balme) and Director, National Endowment for the Humanities (NEH) Research Conference on Philosophical Issues in Aristotle's Biology, Williams College, sponsored by the Council for Philosophical Studies.

1985: Co-director (with G. E. R. Lloyd), Cambridge/Trenton Conference on Aristotle's Philosophy of Biology, King's College, Cambridge.

1986: Director, Trenton/Cambridge Conference on Matter and Explanation in Aristotle's Scientific and Philosophical Works, Trenton State College.

1987: Co-organizer (with D. Devereux and P. Pellegrin), Joint US–France Seminar on Interconnections of Biology, Scientific Method, and

Metaphysics in Scientific and Philosophical Writings of Aristotle, NSF and CNRS (Paris), Ile d'Oléron, France.

1988: Co-director (with J. M. Cooper and M. Frede). Six-week NEH Summer Institute on Aristotle's Metaphysics, Biology, and Ethics, sponsored by the Council for Philosophical Studies.

1995: US Consultant (for W. Kullmann and S. Föllinger, organizers), Symposium on Aristotle's Biology, Bad Homburg, Germany.

2005: Organizer, Workshop on Aristotle's *Generation of Animals*, University of Pittsburgh.

Organizer, Workshop on Ayn Rand as Aristotelian, University of Pittsburgh.

2007: Co-organizer (with Devin Henry), Second Pittsburgh–London (Ont.) Workshop on Aristotle's *Generation of Animals*, University of Pittsburgh.

Co-organizer (with Devin Henry), Third Pittsburgh–London (Ont.), Workshop on Aristotle's *Generation of Animals*, University of Pittsburgh.

2008: Co-organizer (with Devin Henry), Fourth Pittsburgh–London (Ont.), Workshop on Aristotle's *Generation of Animals*, University of Pittsburgh.

2010: Co-organizer (with Devin Henry), Fifth Pittsburgh–London (Ont.) Workshop on Aristotle's *Generation of Animals*, University of Western Ontario.

References

Ackrill, J. L. 1972/3, "Aristotle's Definitions of Psyche," *Proceedings of the Aristotelian Society* 73: 119–33. (Reprinted in Ackrill, J. L. 1997, *Essays on Plato and Aristotle,* Oxford.)

Annas, J. 1999, *Platonic Ethics Old and New,* Ithaca and London.

Balme, D. M. 1987, "Aristotle's Use of Division and Differentiae," in Gotthelf and Lennox 1987, 69–89.

1992, *Aristotle*: De partibus animalium *I and* De generatione animalium *I (with passages from II. 1–3),* Oxford.

Barnes, J. 1981, "Proof and the Syllogism," in Berti 1981, 17–59.

(ed.) 1984. *The Complete Works of Aristotle: The Revised Oxford Translation,* Princeton.

1994, *Aristotle*: Posterior Analytics, 2nd edn., Oxford.

Berti, E. (ed.) 1981, *Aristotle on Science: The "Posterior Analytics,"* Padua.

Bleier, R. 1984, *Science and Gender: A Critique of Biology and its Theories on Women,* New York.

Bodéus, R. 1982, *Le philosophe et la cité,* Paris.

1993, *The Political Dimensions of Aristotle's Ethics,* trans. J. E. Garrett, Albany.

2000, *Aristotle and the Theology of the Living Immortals,* Albany. (Original French edn.: Bodéüs, R. 1992, *Aristote et la théologie des vivants immortels,* Paris and Montreal.)

Bodnár, I. 2005, "Teleology across Natures," *Rhizai* 2: 9–29.

Bolton, R. 1993, "Division, définition et essence dans la science aristotélicienne," *Revue Philosophique de la France et de l'Étranger* 2: 197–222.

1995, "Science and the Science of Substance in Aristotle's *Metaphysics* Z," *Pacific Philosophical Quarterly* 76: 419–69.

2002, "Substance and the Definition of Definition in Aristotle," in Canto-Sperber and Pellegrin 2002, 155–81.

Bostock, D. 1994, *Aristotle's* Metaphysics, *Books Z and H,* Oxford.

Broadie, S. 1990, "Nature and Craft in Aristotelian Teleology," in Devereux and Pellegrin 1990, 389–403.

Broadie, S. and C. Rowe 2002, *Aristotle:* Nicomachean Ethics, *Translation, Introduction, and Commentary,* Oxford.

Brunschwig, J. 1981, "L'objet et la structure des *Seconds Analytiques* d'après Aristote," in Berti 1981, 61–96.

Burnet, J. 1900, *Aristotle:* Nicomachean Ethics, London.
Burnyeat, M. F. 1984, *Notes on Eta and Theta of Aristotle's* Metaphysics, Oxford.
1990, *The* Theatetus *of Plato*, trans. M. J. Levett, revised by M. F. Burnyeat, Indianapolis.
2001, *A Map of* Metaphysics *Zeta*, Pittsburgh.
2004, "Introduction: Aristotle on the Foundations of Sublunary Physics," in de Haas and Mansfeld 2004, 7–24.
2008, "*Kinēsis* vs. *Energeia*: A Much-Read Passage in (but not of) Aristotle's *Metaphysics*," *Oxford Studies in Ancient Philosophy* 34: 219–92.
Canto-Sperber, M. and P. Pellegrin (eds.) 2002, *Le style de la pensée*, Paris.
Charles, D. 1984, *Aristotle's Philosophy of Action*, London.
1986, "Aristotle: Ontology and Moral Reasoning," *Oxford Studies in Ancient Philosophy* 4: 119–44.
1988, "Aristotle on Hypothetical Necessity and Irreducibility," *Pacific Philosophical Quarterly* 69: 1–53.
1991, "Teleological Causation in the *Physics*," in *Aristotle's Physics*, ed. L. Judson, Oxford, 101–28.
1994, "Matter and Form: Unity, Persistence, and Identity," in Scaltsas, Charles, and Gill 1994, 75–105.
2000, *Aristotle on Meaning and Essence*, Oxford.
2004, "GC I.5: Simple Genesis and Prime Matter," in de Haas and Mansfeld 2004, 151–69.
Christ, W. (ed.) 1885, *Aristotelis* Metaphysica, Leipzig.
Code, A. 1996, "Owen on the Development of Aristotle's *Metaphysics*," in *Aristotle's Philosophical Development*, ed. W. Wians, Lanham, 303–26.
1997, "The Priority of Final Causes over Efficient Causes in Aristotle's *PA*," in Kullmann and Föllinger 1997, 127–43.
Colli, G. 1955, *Aristotele, Organon*, Turin.
Cooper, J. M. 1982, "Aristotle on Natural Teleology," in *Language and Logos*, ed. M. Schofield and M. Nussbaum, Cambridge, 197–222.
1999, *Reason and Emotion*, Princeton.
2004, *Knowledge, Nature, and the Good*, Princeton and Oxford.
2009, "*Nicomachean Ethics* VII 1–2: Introduction, Method, and Puzzles," in *Aristotle's* Nicomachean Ethics, *Book VII. Symposium Aristotelicum*, ed. Carlo Natali, Oxford, 9–39.
de Haas, F. and J. Mansfeld (eds.) 2004, *Aristotle's* On Generation and Corruption, *Book I. Symposium Aristotelicum*, Oxford.
Denniston, J. D. 1966, *The Greek Particles*, 2nd edn. Oxford.
Deslauriers, M. 1998, "Sex and Essence in Aristotle's Metaphysics and Biology," in *Feminist Interpretations of Aristotle*, ed. Cynthia Freeland, University Park, 138–67.
Devereux, D. and P. Pellegrin (eds.) 1990, *Biologie, Logique et Métaphysique chez Aristote*, Paris.
Diels, H. and W. Kranz 1951, *Die Fragmente der Vorsokratiker*, 3 vols., 6th edn., Dublin and Zurich.

Duke, E. A., Hicken, W. F., Nicoll, W. S. M., Robinson, D. B., and J. C. G. Strachan (eds.) 1995, *Platonis opera*, vol. 1, Oxford.

Düring, I. 1943, *Aristotle's* De partibus animalium*: Critical and Literary Commentaries*, Göteborg.

Elshtain, J. B. 1981, *Public Man, Private Woman*, Princeton.

Falcon, A. 2005, *Aristotle and the Science of Nature*, Cambridge.

Frede, M. 1987, "Categories in Aristotle," *Essays in Ancient Greek Philosophy*, Minneapolis, 29–48.

1990, "The Definition of Sensible Substances in *Metaphysics* Z," in Devereux and Pellegrin 1990, 113–29.

1994, "Potentiality in the *Metaphysics*," in Scaltsas, Charles, and Gill 1994, 188–93.

Frede, M. and G. Patzig 1988, *Aristoteles Metaphysik Z*, Munich.

Freeland, C. 1987, "Aristotle on Bodies, Matter, and Potentiality," in Gotthelf and Lennox 1987, 392–407.

Furley, D. 1985, "The Rainfall Example in *Physics* ii.8," in Gotthelf 1985a, 177–82.

Gerson, L. P. 2005, *Aristotle and Other Platonists*, Ithaca and London.

Gill, M. L. 1989, *Aristotle on Substance: The Paradox of Unity*, Princeton.

1995, "Aristotle on Substance and Predication," *Ancient Philosophy* 15: 511–20.

2003, "The Unity of Substances in *Metaphysics* H.6," (in Portuguese) in *Cadernos de História e Filosofia da Ciência*, Series 3, 13: 177–203.

2008, "Form–Matter Predication in *Metaphysics* Θ.7," in *Dunamis: Autour de la puissance chez Aristote*, ed. M. Crubelier, A. Jaulin, and D. Lefebvre, Leuven, 391–427.

2009, "The Divine Method in Plato's *Philebus*," in *Plato's* Philebus*. Selected Papers from the Eighth Symposium Platonicum*, ed. J. Dillon and L. Brisson, Sankt Augustin, 36–46.

Goldin, O. 1996, *Explaining an Eclipse. Aristotle's* Posterior Analytics *2.1–10*, Ann Arbor.

Gotthelf, A. (ed.) 1985a, *Aristotle on Nature and Living Things*, Pittsburgh and Bristol.

1985b, "Notes towards a Study of Substance and Essence in Aristotle's *Parts of Animals* ii–iv," in Gotthelf 1985a, 27–54.

1987, "First Principles in Aristotle's *Parts of Animals*," in Gotthelf and Lennox 1987, 167–98.

1989, "The Place of the Good in Aristotle's Natural Teleology," in *Proceedings of the Boston Area Colloquium in Ancient Philosophy* vol. IV, ed. J. J. Cleary and D. Shartin, Lanham, MD, 113–39.

1997a, "Division and Explanation in Aristotle's *Parts of Animals*," in *Beiträge zur antiken Philosophie. Festschrift für Wolfgang Kullmann*, ed. Hans-Christian Günther and A. Rengakos, Stuttgart, 215–30.

1997b, "The Elephant's Nose: Further Reflections on the Axiomatic Structure of Biological Explanation in Aristotle," in Kullmann and Föllinger 1997, 85–96.

1997c, "Understanding Aristotle's Teleology," in *Final Causality in Nature and Human Affairs*, ed. R. F. Hassling, Washington, 71–85.
Gotthelf, A. and J. G. Lennox (eds.) 1987, *Philosophical Issues in Aristotle's Biology*, Cambridge.
Gregoric, P. 2005, "Plato's and Aristotle's Explanation of Human Posture," *Rhizai* 2: 183–96.
Halper, E. 1989, *One and Many in Aristotle's* Metaphysics, Columbus.
Harte, V. 1996, "Aristotle's *Metaphysics* H6: a Dialectic with Platonism," *Phronesis* 41: 276–304.
Hayduck, M. (ed.), 1981, *Alexander: Aristotelis Metaphysica commentaria*, *Commentaria in Aristotelem Graeca*, vol. 1, Berlin.
Henry, D. 2006a, "Aristotle on the Mechanism of Inheritance," *Journal for the History of Biology* 39: 425–55.
2006b, "Understanding Aristotle's Reproductive Hylomorphism," *Apeiron* 39.3: 269–300.
Irwin, T. H. 1990, "The Good of Political Activity," in *Aristoteles' Politik*, ed. G. Patzig, Göttingen, 73–98.
Jaeger, W. (ed.) 1957, *Aristotelis* Metaphysica, Oxford.
Johnson, M. R. 2005, *Aristotle on Teleology*, Oxford.
Judson, L. 1991, "Chance and 'Always or For the Most Part,'" in *Aristotle's Physics*, ed. L. Judson, Oxford, 73–99.
2005, "Aristotelian Teleology," *Oxford Studies in Ancient Philosophy* 29: 341–66.
Kahn, C. H. 1985, "The Place of the Prime Mover in Aristotle's Teleology," in Gotthelf 1985a, 183–205.
Keyt, D. 1991, "Three Basic Theorems in Aristotle's *Politics*," in *A Companion to Aristotle's* Politics, ed. D. Keyt and F. Miller, London, 54–79.
Kosman, A. 1984, "Substance, Being and *Energeia*: The Argument of *Metaphysics* Theta," *Oxford Studies in Ancient Philosophy* 2: 121–49.
1987, "Animals and Other Beings in Aristotle," in Gotthelf and Lennox 1987, 360–91.
Kullmann, W. and S. Föllinger (eds.) 1997, *Aristotelische Biologie: Intentionen, Methoden, Ergebnisse*, Stuttgart.
Lamb, C. 1823, "A Dissertation upon Roast Pig," *Essays of Elia*, London.
Lear, G. R. 2004, *Happy Lives and the Highest Good*, Princeton.
Lee, Mi-Kyoung 2005, *Epistemology after Protagoras*, Oxford.
Lennox, J. G. 1985, "Theophrastus on the Limits of Teleology," in *Theophrastus of Eresus: On his Life and Works*, ed. W. W. Fortenbaugh, P. M. Huby, and A. A. Long, New Brunswick, NJ and Oxford, 143–64. (Reprinted in Lennox 2001b, 259–79.)
1987a, "Divide and Explain: the *Posterior Analytics* in Practice," in Gotthelf and Lennox 1987, 90–119. (Reprinted in Lennox 2001b, 7–32.)
1987b, "Kinds, Forms of Kinds, and the More and the Less in Aristotle's Biology," in Gotthelf and Lennox 1987, 339–59. (Reprinted in Lennox 2001b, 160–81.)

1996. "Putting Philosophy of Science to the Test: The Case of Aristotle's Biology," in *PSA Proceedings: 1994*, vol. II, ed. M. Forbes, D. Hull, and R. Burian, East Lansing, 239–47.
2001a, *Aristotle:* On the Parts of Animals *I–IV*, Oxford.
2001b, *Aristotle's Philosophy of Biology: Studies in the Origins of Life Science*, Cambridge.
2005a, "Getting a Science Going: Aristotle on Entry Level Kinds," in Wolters and Carrier 2005, 87–100.
2005b, "The Place of Zoology in Aristotle's Natural Philosophy," in Sharples 2005a, 58–70.
2010, "Bios and Explanatory Unity in Aristotle's Biology," in *Definition in Greek Philosophy*, ed. D. Charles, Oxford, chapter 9.
Lewis, F. 1991, *Substance and Predication in Aristotle*, Cambridge.
1994, "Aristotle on the Relation between a Thing and its Matter," in Scaltsas, Charles, and Gill 1994, 247–77.
1995, "Aristotle on the Unity of Substance," *Pacific Philosophical Quarterly* 76: 222–65. (Reprinted in Lewis, F. A. and R. Bolton [eds.] 1996, *Form, Matter, and Mixture in Aristotle*, Oxford, 39–81.)
Liddell, H. G., Scott, R., and H. S. Jones 1940, *A Greek–English Lexicon*, Oxford.
1996, *A Greek–English Lexicon*, 9th edn. revised, Oxford.
Lloyd, G. E. R. 1966, *Polarity and Analogy*, Cambridge.
1990, "Aristotle's Zoology and his Metaphysics," in Devereux and Pellegrin 1990, 7–35.
1993, "L'idée de la nature dans la *Politique* d'Aristote," in *Aristote*, Politique, ed. P. Aubenque, Paris, 135–59. (Eng. trans. in Lloyd, G. E. R. 1996, *Aristotelian Investigations*, Cambridge, 184–204.)
Loux, M. J. 1995a, "APA Symposium on Aristotle's *Metaphysics*," *Ancient Philosophy* 15: 495–510.
1995b, "Composition and Unity: An Examination of *Metaphysics* H.6," in *The Crossroads of Norm and Nature: Essays on Aristotle's Ethics and Metaphysics*, ed. May Sim, Lanham, 242–79.
Maier, H. 1896–1900, *Die Syllogistik des Aristoteles*, 3 vols., Tübingen.
Makin, S. 2003, "What Does Aristotle Mean by Priority in Substance?," *Oxford Studies in Ancient Philosophy* 34: 209–38.
2006, *Aristotle:* Metaphysics *Book Θ*, Oxford.
Mansion, S. 1976, *Le jugement d'existence chez Aristote*, 2nd edn., Louvain.
Matthen, M. 2001, "The Holistic Presuppositions of Aristotle's Cosmology," *Oxford Studies in Ancient Philosophy* 20: 171–99.
Mayhew, R. 2004, *The Female in Aristotle's Biology: Reason or Rationalization*, Chicago.
McDowell, J. 1973, *Plato's* Theaetetus, Oxford.
1979, "Virtue and Reason," *The Monist* 72: 331–50.
McKirahan, R. D. 1992, *Principles and Proofs: Aristotle's Theory of Demonstrative Science*, Princeton.

Mignucci, M. 1975, *L'argomentazione dimostrativa in Aristotele: commento agli Analytici Secondi*, vol. 1, Padua.
Modrak, D. 2001, *Aristotle's Theory of Language and Meaning*, Cambridge.
2007, "Form and Function," *Proceedings of the Boston Area Colloquium in Ancient Philosophy* 22: 111–35.
Mure, G. R. G. 1928, Analytica posteriora: *The Works of Aristotle Translated into English*, Oxford.
Osborne, C. 1988, "Topography in the *Timaeus*: Plato and Augustine on Mankind's Place in the Natural World," *Proceedings of the Cambridge Philological Society* 214: 104–14.
Peck, A. L. (trans.) 1963, *Aristotle:* Generation of Animals, London and Cambridge, MA.
Pellegrin, P. 1982, *La classification des Animaux chez Aristote*, Paris.
1985, "Zoology without Species," in Gotthelf 1985a, 95–115.
2002, "Les ruses de la nature et l'éternité du mouvement. Encore quelques remarques sur la finalité chez Aristote," in Canto-Sperber and Pellegrin 2002, 296–323.
2005, *Aristote: Seconds Analytiques Organon IV*, Paris.
Peramatzis, M. 2008, "Aristotle's Notion of Priority in Nature and Substance," *Oxford Studies in Ancient Philosophy* 35: 187–247.
Reeve, C. D. C. 1998, *Aristotle:* Politics. Indianapolis and Cambridge.
2002, *Substantial Knowledge*, Indianapolis.
Reydams-Schils, G. 1997, "Plato's World Soul: Grasping Sensibles without Sense-perception," in *Interpreting the* Timaeus–Critias, ed. T. Calvo and L. Brisson, Sankt Augustin, 261–5.
Rhenius, R. 2005, *Die Einheit der Substanzen bei Aristoteles*, Berlin.
Romeyer Dherbey, G. 1996, *Corps et Ame: Sur le* De Anima *d'Aristote*, Paris.
Rorty, R. 1973, "Genus as Matter: A Reading of *Metaphysics* Z–H," in *Exegesis and Argument* (Suppl. vol. 1 of *Phronesis*), ed. E. N. Lee, A. P. D. Mourelatos, and R. M. Rorty, New York, 393–420.
Ross, W. D. 1924, *Aristotle's* Metaphysics, 2 vols., Oxford.
1928, *Metaphysica. The Works of Aristotle Translated into English*, Oxford.
1949, *Aristotle's* Prior *and* Posterior Analytics, Oxford.
Scaltsas, T. 1994, "Substantial Holism," in Scaltsas, Charles, and Gill 1994, 107–28.
Scaltsas, T., Charles, D., and M. L. Gill (eds.) 1994, *Unity, Identity and Explanation in Aristotle's Metaphysics*, Oxford.
Sedley, D. 1991, "Is Aristotle's Teleology Anthropocentric?," *Phronesis* 36: 179–96.
2000a, "The Ideal of Godlikeness," in *Plato*, vol. II, ed. G. Fine, Oxford, 309–28.
2000b, "*Metaphysics* Lambda 10," in *Aristotle's* Metaphysics *Lambda*, ed. M. Frede and D. Charles, Oxford, 327–50.
2005, "Les origines des preuves stoïciennes de l'existence de dieu," *Revue de Métaphysique et de Morale* 4: 461–87.
2007, *Creationism and its Critics in Antiquity*, Berkeley and Los Angeles.
Sellars, W. 1967, "Aristotle's *Metaphysics*: An Introduction," *Philosophical Perspectives*, Springfield.

Sharples, R. W. 1983, *Alexander of Aphrodisias on Fate*, London.
2003, "Threefold Providence: The History and Background of a Doctrine," in *Ancient Approaches to Plato's Timaeus*, ed. R. W. Sharples and A. Sheppard, London, 107–27.
(ed.) 2005a, *Philosophy and the Sciences in Antiquity*, London.
2005b, "Some Thoughts on Aristotelian Form: With Special Reference to *Metaphysics* Z.8," *Science in Context* 18: 93–109.
Shorey, P. 1924, "The Origin of Syllogism," *Classical Philology* 19: 1–19.
Smyth, H. W. 1984, *Greek Grammar*, Cambridge.
Solmsen, F. 1929, *Die Entwicklung der aristotelischen Logik und Rhetorik*, Berlin.
1963, "Nature as Craftsman in Greek Thought," *Journal of the History of Ideas* 24: 473–96.
Sorabji, R. 1980, *Necessity, Cause and Blame*, London.
Todd, R. B. 1976, "The Four Causes: Aristotle's Exposition and the Ancients," *Journal of the History of Ideas* 37: 319–22.
Tricot, J. 1962, *Les Seconds analytiques*, Paris.
Verdenius, W. J. 1981, "Notes on Some Passages from Book I," in Berti 1981, 343–57.
Villee, C. A. and V. G. Dethier 1971, *Biological Principles and Processes*, Philadelphia.
Wallies, M. (ed.) 1900, *Themistius: In Analyticorum posteriorum paraphrasis*, *Commentaria in Aristotelem Graeca*, vol. v.1, Berlin.
(ed.) 1909, *Philoponus: In Aristotelis Analytica posteriora commentaria cum Anonymo in Librum II*, *Commentaria in Aristotelem Graeca*, vol. xiii.3, Berlin.
Wardy, R. 1993, "Aristotelian Rainfall or the Lore of Averages," *Phronesis* 38: 18–30.
2005, "The Mysterious Aristotelian Olive," *Science in Context* 18: 69–91.
Watson, J. D. 1970, *Molecular Biology of the Gene*, New York.
Witt, C. 1989, *Substance and Essence in Aristotle*, Ithaca.
Wolters, G. and M. Carrier (eds.) 2005, *Homo Sapiens und Homo Faber*, Berlin.
Zabarella, J. 1594, *In duos Aristotelis libros Posteriores analyticos commentarii* (in *Opera Logica*), Bale.

Index locorum

General index